# Atmospheric Electrostatics

**ELECTRONIC & ELECTRICAL ENGINEERING RESEARCH STUDIES**

**ELECTROSTATICS AND ELECTROSTATIC APPLICATIONS SERIES**

*Series Editor:* **Dr. J. F. Hughes,** *Department of Electrical Engineering, University of Southampton, England*

1. Static Elimination
   **T. Horváth** *and* **I. Berta**
2. Electrostatic Hazards in the Petroleum Industry
   **W. M. Bustin** *and* **W. G. Dukek**
3. Electrophotography Principles and Optimization
   **Merlin Scharfe**
4. Electrostatic Powder Coating
   **J. F. Hughes**
5. Electrostatic Mineral Separation
   **Ion I. Inculet**
6. Computer Modelling in Electrostatics
   **D. McAllister, J. R. Smith** *and* **N. J. Diserens**
7. Atmospheric Electrostatics
   **Lars Wåhlin**

# Atmospheric Electrostatics

**Lars Wåhlin**
*Colutron Research Corporation*
*Boulder, Colorado, USA*

**RESEARCH STUDIES PRESS LTD.**
Letchworth, Hertfordshire, England

**JOHN WILEY & SONS INC.**
New York · Chichester · Toronto · Brisbane · Singapore

RESEARCH STUDIES PRESS LTD.
58B Station Road, Letchworth, Herts. SG6 3BE, England



**Marketing and Distribution:**

*Australia, New Zealand, South-east Asia:*
Jacaranda-Wiley Ltd., Jacaranda Press
JOHN WILEY & SONS INC.
GPO Box 859, Brisbane, Queensland 4001, Australia

*Canada:*
JOHN WILEY & SONS CANADA LIMITED
22 Worcester Road, Rexdale, Ontario, Canada

*Europe, Africa:*
JOHN WILEY & SONS LIMITED
Baffins Lane, Chichester, West Sussex, England

*North and South America and the rest of the world:*
JOHN WILEY & SONS INC.
605 Third Avenue, New York, NY 10158, USA

***Library of Congress Cataloging in Publication Data***

Wåhlin, Lars.
Atmospheric electrostatics.

(Electronic & electrical engineering research studies. Electrostatics and electrostatic applications series; 7)
Bibliography: p.
Includes index.
1. Atmospheric electricity. I. Title. II. Series.
QC961.W34 1986 551.5'63 86-13739
ISBN 0 471 91202 6 (Wiley)

***British Library Cataloguing in Publication Data***

Wåhlin, Lars.
Atmospheric electrostatics.——(Electronic & electrical engineering research studies. Electrostatics and electrostatic applications series; 7)
1. Atmospheric electricity
I. Title II. Series
551.5'6 QC961

ISBN 0 86380 042 4
ISBN 0 471 91202 6 Wiley

ISBN 0 86380 042 4 (Research Studies Press Ltd.)
ISBN 0 471 91202 6 (John Wiley & Sons Inc.)

Printed in Great Britain by Short Run Press Ltd., Exeter

# Editorial Foreword

Atmospheric Electrostatics is a welcome addition to the Electrostatics Series of monographs.

The author has skilfully combined fundamental theories of atmospheric electrostatic phenomena with his own unique explanation of thundercloud charging. The arguments put forward are very convincing and are supported by appealingly simple experiments. Mathematical treatment of the subject is kept to a minimum, which enables the non-expert to follow the reasoning with ease. At the same time, the introduction of the electrochemical model for charge exchange makes this compulsory reading for atmospheric scientists who are more familiar with the traditional theories of thundercloud charging.

The subject is presented in a very logical manner, beginning with early 19th Century experiments and leading finally to extraterrestrial lightning phenomena. Congratulations to the author for presenting this complex subject in such a way that even a novice to the field will have no difficulty in following the text. All levels of readers are guaranteed a fascinating conducted tour through this most fundamental of electrical phenomena.

Dr. J.F.Hughes,PhD, CEng, FIEE, FInstP.
Southampton. April 1986

# Preface

Static electricity became a fashionable science in the early 16th century and several investigators drew a parallel between the sparks produced in the laboratories and that of lightning and thunder produced during foul weather. It is surprising, however, that even today with our highly advanced technology in electronics and space science, we still do not know what causes thunderclouds to charge. The purpose of this book is to give a general summary of atmospheric electricity and to discuss several proposed charging mechanisms, including recent important discoveries in atmospheric electrochemistry. Atmospheric electrochemistry becomes important when we realize that the atmosphere, due to the constant bombardment of cosmic rays, is ionized and behaves very much like an electrolyte. Electrochemical potentials are produced on material surfaces that are exposed to our ionized atmosphere and are as common as contact potentials generated when dissimilar materials touch each other.

This book is not a review of the most current publications on atmospheric electricity but serves as an overview of the basic problems still at large and the purpose is to try to inspire fresh blood into the oldest field of electricity. Two excellent textbooks are recommended for those interested in a detailed picture of the electrical structure of our atmosphere: H. Israel, *Atmospheric Electricity* Vol. 1, (1970), and Vol.2

(1973); J.A. Chalmers, *Atmospheric Electricity*, (1957).

The author wishes to express thanks to the Burndy Library; The High Voltage Research Institute, Uppsala University, Sweden, and the High Voltage Laboratory, T.U. Munich for supplying historic illustrations.

Thanks are also due to Dr. John F. Hughes for initiating this work and for editing and facilitating the publication of this book.

Boulder, April 1986. Lars Wåhlin

# Table of Contents

# List of Illustrations

# CHAPTER 1
# Historical Background

## 1.1 PRIMITIVE BELIEFS

An electric storm is one of nature's most spectacular phenomena, and its display of lightning and thunder has fascinated and frightened man throughout time. In ancient times it was believed that the great gods were responsible for hurling thunderbolts that could kill, upturn boulders, split trees or kindle fires. Man's early steps towards civilization began when he learned how to control fires started by lightning and use it against predatory animals. Tales from American Indians explain how the world was a cold place before the first fire was started by the thunderbird god. As recently as the 1870's German soldiers were convinced that magic power from the **donnerkeil**, or lightning stone, would protect them from French bullets. Lightning stones, believed to be spearheads of lightning bolts, could be found buried where lightning had struck. The lightning stones and their fragments were sold throughout Europe for many hundreds of years and were thought to protect from illness and evil. We know now that most of the stones and fragments found were old relics and artifacts from the Stone Age (Lundquist 1969).

Today we smile at old superstitions, yet we too bend to the mystery of lightning and its atmospheric implications, for modern man has yet to understand it.

## 1.2 EARLY ELECTROSTATICS

The first person on record to have suggested a relationship between electricity and lightning was an Englishman named D. William Wall (1708). He noted a similarity between lightning and the crackling sparks produced by the rubbing of amber. S. Gray (1735) and A.G. Rosenberg (1745) both mention the similarity between lightning phenomena and electric fire produced by electricity machines in the laboratory; and in a book published in Leipzig in 1746, J.H. Winkler describes several resemblances between lightning and electricity. During this time improved electricity machines and Leyden jars became readily available and a new era of electrical science was born. Many more scientists, among them Benjamin Franklin, also questioned the nature of lightning, recognizing its similarity to the snapping sparks produced in the laboratory. "How loud must be the crack of 10,000 acres of electrified cloud!" exclaimed Franklin. In a letter to Dr. John Mitchel of the Royal Society in England, he enclosed a treatise, "The Sameness of Lightning and Electricity". According to Mitchel, the paper was read by the Society amidst laughter from its professed experts on electricity. Another paper, dated July 1750, was sent to the Royal Society through a friend, a Mr. Collinson. In this paper Franklin described how electric "fluid" is attracted to pointed conductors. "Might not the same principles be of use to man in teaching him to protect houses, churches, ships and other structures from damage occasioned by lightning?" he asked. Thus the idea of the lightning rod was born.

As yet no experimental tests had been performed to prove that lightning was an electrical phenomenon. Franklin therefore proposed an experiment to answer the question once and for all. On a high tower or steeple a sentry-box was to be erected large enough to contain a man and an electrically-insulating platform. A long pointed rod or antenna would be attached to the sentry-box by means of insulators and would be connected to the insulated platform inside (Fig. 1). A man standing on the insulated platform would thus become charged from the rod when a thundercloud passed overhead. At will he might then draw sparks from his fingertips to the surrounding grounded wall.

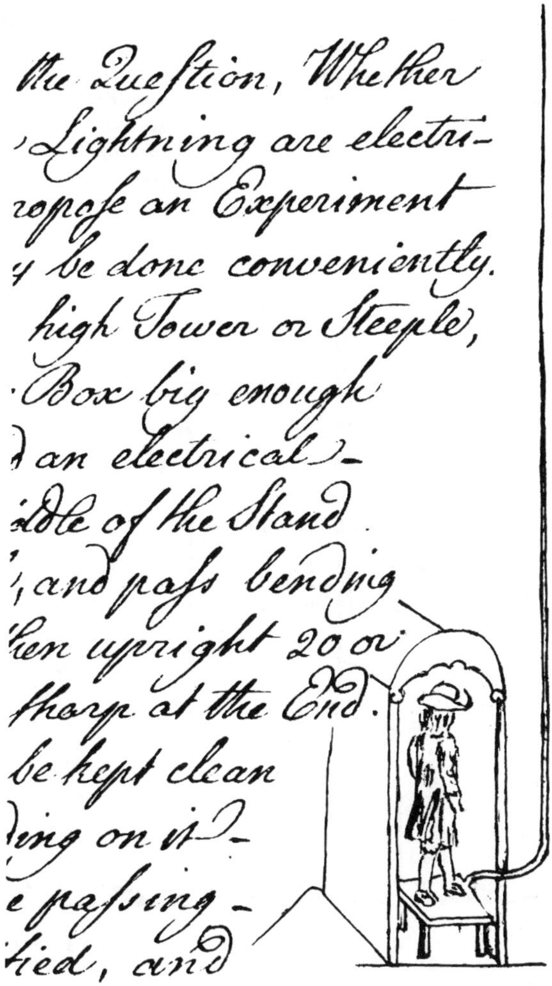

Fig. 1 Franklins's proposed sentry-box experiment.

In principle it was just another lightning rod. From laboratory experiments Franklin knew that charge could be drawn from a nearby charged body by means of a pointed rod, which attracted electric "fire".

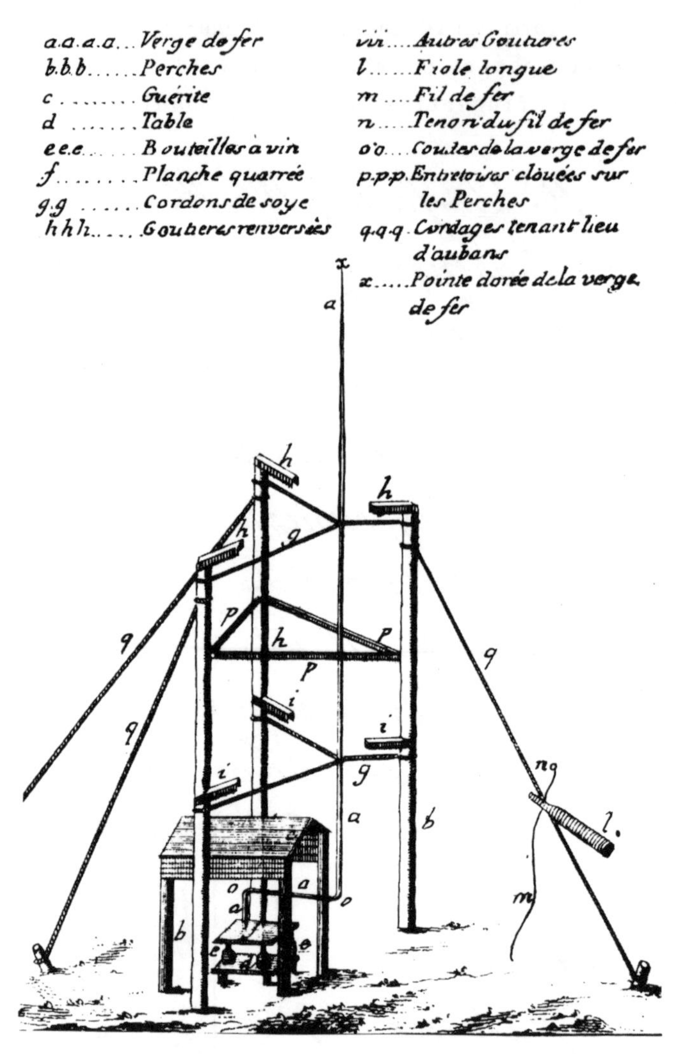


Fig. 2 D'Alibard's version of Franklin's sentry-box experiment.

The experiment was never to be performed by Franklin, due to the lack of financial aid, yet with time Franklin's ideas gained increasing approval from the Royal Society. In 1751 several of his papers were published in England in book form and soon thereafter translated into French by the naturalist

D'Alibard. So intrigued was D'Alibard by the sentry-box experiment that he decided to put it to the test himself. An experimental structure slightly different in design was erected outside Paris at Marly. (Fig. 2). By the 10th of May, 1752 D'Alibard (1752) had successfully determined that thunderclouds are indeed electrically charged.

In America a few weeks later Franklin, unaware of D'Alibard's success, performed his famous kite experiment. It was a poor man's experiment, simple and brilliant. It demonstrated that lack of financial help is an insufficient deterrent to genius. (One wonders if Dr. Franklin was advised, when looking for financial support, to go fly a kite.) In his kite experiment Franklin not only confirmed the electrical character of lightning but also, more importantly, found clouds to be negatively charged at the base and positively charged on top, thus forming giant electric dipoles in our atmosphere.

### 1.3 FRANKLIN VERSA NOLLET

It is interesting to note that Franklin spoke of positive and negative electricity. This is a product of his own theory on electricity and the concept of positive and negative charge is still being used today. Franklin envisaged electricity as a kind of fluid where a body could have either an excess or a deficit of fluid. For example, if a person standing on an insulated stool were to touch the glass cylinder of an electricity machine he would lose some of the fluid. A deficit or loss of fluid represents a minus state and excess fluid equals a plus state. The sum of positive and negative electricity is always nil. The plus state of the fluid was determined by Franklin as follows: A positively charged pith ball would snuff out the flame on a candle *i.e.* the flame will retreat from the ball while the ball is trying to rid itself of excess fluid. A modern explanation is that the flame, which contains a great number of positive ions produced by the heat and combustion, will be repelled by any other positive charge in its vicinity. One consequence of Franklin's theory is the convention of labelling the direction of electric current. Although in metals the charge is carried by electrons flowing from minus to plus, we are now forced to assume that electric current or charge is flowing from plus to minus in order to keep things algebraically consistent.

The Abbé Nollet (1700-1770) of Paris was the authority on electricity in Europe during the time of Franklin's discoveries. Nollet was aware of two different kinds of electricity, namely glass electricity and resin electricity. He believed that the two different kinds of electricity were liquid streams flowing in and out of electrified bodies in opposite directions, *effluence* and *affluence*. Nollet maintained a very negative attitude towards Franklin's ideas on electricity and especially to the lightning rod. This is perhaps one reason why it took so long for the lightning rod to become established in Europe. The Abbé Nollet, however, was a master of experimental electricity. He often performed electricity demonstrations before Louis XV and his court. A typical demonstration would be to pass a high voltage static charge from a Leyden jar, through seven hundred monks holding hands, causing them to jump in perfect unison, to the King's delight. Nollet was often criticized for activities such as this. His theories on electricity were eventually abandoned in favour of Franklin's ideas which, to a certain extent, still are used today.

### 1.4 THE LIGHTNING ROD

A year after the famous tests of D'Alibard and Franklin a Russian professor named Richmann attempted to repeat the sentry-box experiment. The result is pictured in Fig. 3. Professor Richmann was killed instantly when lightning struck his antenna on the roof of the laboratory in St. Petersburg. As a result of this accident the general public grew skeptical of the lightning rod as a protective device.

**Fig. 3 Professor Richmann is killed when lightning strikes his experimental antenna in St. Petersburg, 1753.**

Did not the death of Professor Richmann prove that lightning rods attract lightning? Some twenty years later Europeans finally accepted Franklin's invention, but the arguments against it were many. One such argument, advanced by a number of scientists spearheaded by Nollet, ran as follows:

If a large structure like a church steeple is not

spared by lightning, how much larger must the lightning rod itself be to withstand the devastating power of a severe bolt?

The argument seemed valid at the time, for church steeples literally exploded when struck by lightning (Ohm's Law had not yet been formulated). The resistance of a church steeple is perhaps one megohm, and since the average lightning bolt carries about 25,000 amperes, it will dissipate a peak power of $R \times I^2 = 6.25 \times 10^8$ megawatts. Although the duration of the lightning bolt is short, its total energy is considerable. The function of the lightning rod is, of course, to lower the resistance across the structure a million or more times and thus to permit the energy to dissipate into the ground.

**Fig. 4 Hat pin and umbrella lightning rods (Paris, 1778).**

Eventually the lightning rod became a common sight on both sides of the Atlantic. Europeans, who had once strongly rejected the notion, embraced it with an enthusiasm that led to extravagant measures for protecting life and limb from the perils of lightning (Fig. 4).

Ironically, Europeans today are occasionally amused by the different shapes and designs of American lightning rods which they feel are costly and only of psychological importance, see Fig. 5. (Müller-Hillebrand 1963).

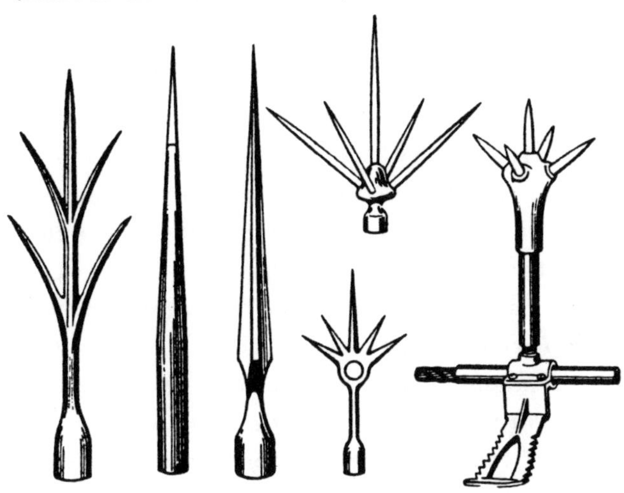

**Fig. 5 Points according to American standards (1959) not cheap and only of psychological importance.**

## 1.5 LATER DEVELOPMENTS

The D'Alibard-Franklin experiments were repeated by many investigators and most noteworthy is perhaps L. Lemonnier (1752) who, with his more sensitive apparatus, discovered that weak electrical charges could be detected in the atmosphere in the absence of clouds. He also noticed a difference in electric intensity during night and day. The discovery of Lemonnier is important because it gave birth to a new field of research in atmospheric physics, namely "Fairweather Electricity".

In 1775 the Italian scientist Beccaria (1775) confirmed Lemonnier's findings with one important addition, that pointed rods and insulated wires suspended in the atmosphere became positively charged relative to the earth's surface during fine weather, under cloud-free skies. This meant that an electric field is present in the atmosphere even when no thunderclouds are near. Beccaria also noticed a polarity reversal when a thundercloud passed overhead, indicating the introduction of negative charge from above. Thus he substantiated Franklin's earlier finding that thunderclouds are generally charged negatively at the bottom and positively at the top, an observation that can readily be confirmed today.

The first notable attempt to explain the origin of fairweather electricity was made by Lord Kelvin (1860). He was first to envisage the fairweather field as electric field lines which must stretch from the earth's surface and terminate on charges in the atmosphere. The direction of the field is towards the earth's surface. He theorized that the charged atmosphere forms a giant capacitor with the earth's surface. The capacitor model was later expanded by Kennely and Heavyside (1902) who suggested that the conducting layer in the atmosphere is the ionosphere which forms a spherical capacitor with the earth's surface where each "plate" of the capacitor carries a charge of nearly one million coulombs. A charge of this magnitude will, of course, produce an electric field of a few hundred volts per metre at the earth's surface. This is the normal value of the persisting fairweather field and Lord Kelvin, not knowing what process had charged the capacitor plates, at least seemed to have a reasonable idea explaining the structure of the fairweather field. Unfortunately for the attractive theory, later discovery of ions and cosmic rays proved the atmosphere to be electrically conducting. Lord Kelvin's capacitor model would therefore lose its charge in less than 10 minutes at a rate of nearly 2000 coulombs per second (=2000 amperes). It should be mentioned that C.A. Coulomb, the father of the electrostatic force law, discovered in 1785 that air is slightly conducting, an observation that was not understood at the time.

Where does the atmospheric fairweather charge come from? How is it replenished at the rate of 2000 coulombs per second? Many answers have been proposed, the most popular being that of C.T.R. Wilson in 1925. Wilson suggested that all thunderstorms around the world are electrical generators which by their violent discharges continually supply electricity to the earth-ionosphere system. More recent evidence, however, weighs heavily against this evocative idea. Insufficient charge is available from worldwide thunderstorms to drive such a global electric circuit. Furthermore, data presented by Imyanitov and Chubarina, among others, demonstrate that annual variations of the fairweather field are not in phase with typical thunderstorm activity throughout the world for the same period. Despite the arguments against Wilson's proposal and the lack of evidence to support it, many investigators still favour the idea since, until very recently, there has been no other explanation available. Now there is a new theory in competition with Wilson's concept based on the electrochemical properties of the atmosphere. This new theory is discussed in Chapter 2.

Serious research on lightning and thunderclouds started late. With the birth of electric power in the early 20th century and the many power failures due to lightning, a better understanding of lightning and thundercloud charging was necessary. Charles Proteus Steinmetz, a German immigrant to America who worked as an engineer at General Electric Laboratories to develop lightning arrestors, might be considered as one of the early pioneers of modern lightning research. His work led to the construction of high-power high-voltage generators which could simulate lightning flashes. The machines consisted primarily of a large capacitor bank which was charged by a high voltage transformer via rectifiers or diodes. For the first time research could be performed on large electric discharges under controlled conditions. Much information was gathered by Steinmetz and his lightning machine which has helped us understand electric transients and has aided in the design of lightning protectors. The invention of the oscilloscope increased our knowledge on the lightning discharge because of its fast response to electrical transients. H. Norinder at the University of Uppsala High Voltage Research Institute was first to obtain such oscillograms from lightning surges in 1925. Parallel with electric research during the first part of the 20th century was the development of a high-speed camera which would record the optical properties of lightning. The first camera suitable for recording the rapid changes in lightning flashes was created by Sir Charles Boys who himself, unfortunately, did not obtain any satisfactory pictures of lightning flashes. It was not until 1933 that the Boys camera became a main contributor to what we now know about lightning discharges. It was B.F.J. Schonland and his team in South Africa who discovered the different sequences of a lightning flash with the aid of a Boys camera (Fig. 6) and revealed the

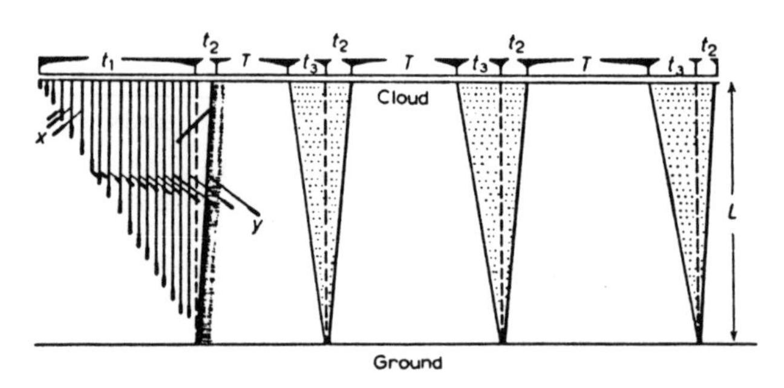


**Fig. 6 Boys camera photograph of cloud to ground discharge.**

initial process of the stepped leader. Boys' pictures show that a bright-tipped leader works its way down from the cloud in steps. When the leader gets near ground it is met by the main return stroke which carries the main discharge current through the ionized conducting path provided by the leader. Normally the stepped leader is invisible to the naked eye and only the main stroke can be seen.

At the present time much is still unknown about atmospheric electricity and its origin. What mechanism is responsible for the build up of cloud charges and what generates the electric fairweather field? How is charge drained from a cloud or how does the lightning bolt connect up to all the myriads of charged drops in the cloud? These are still valid questions and, in the author's opinion, is the reason why Atmospheric Electricity is one of the most challenging fields in science today.

# CHAPTER 2

# The Electric Structure of the Atmosphere

## 2.1 IONS

The existence of ions in the atmosphere is the fundamental reason for atmospheric electricity. An absence of ions would mean zero electric field in the atmosphere and most probably no thunderstorms or lightning. The concept of positive and negative ions as charge carriers in the atmosphere was first put forward by J. Elster and H. Geitel (1899) in order to explain the electric conductivity of air. Much work has since been done on ions and their role in atmospheric electric phenomena. Today we know there are mainly three classes of ions, namely small ions, intermediate ions and large ions. Most important are the small ions since their higher mobility allows them to take a more active part in the transfer of charge throughout the atmosphere. The mobility of ions can be measured in metres per second per volt per metre which signifies the velocity that an ion will reach when subjected to an electric field of one volt per metre. For small ions the mobility is of the order of 0.0001 with a slight edge of the negative ion over its positive counterpart. In fact, the negative to positive mobility ratio of small ions is about 1.25 (Wåhlin 1985) which is a paradox since negative ions are believed to be more massive than positive ions. One explanation (Papoular, 1965) is that for part of its lifetime a negative ion is really an electron jumping from molecule to molecule. Molecules such as NO and $NO_2$ are believed to dominate the negative small ion population while oxonium and water might make up the positive small ions in the atmosphere. Their true molecular structure and mass are not well known because it is difficult to get spectroscopic mass analysis of small ions in the lower atmosphere. The problem is their relative short life time, about 100 seconds, which is much shorter than the transit time required for molecules or ions to reach the source end of a mass analyser.

The ionization in the lower atmosphere is mostly caused by cosmic rays and natural radioactivity. Ions are also produced in and near thunderclouds by lightning and corona processes. Cosmic rays originate from solar flares and other galactic objects such as supernovas and exploding stars. One interesting thought is; do stellar events affect our lives here on earth? We know that cosmic rays are by far the major ion producers in the lower atmosphere and if thunderstorms need ions to feed on in order to charge, we certainly would not have thunderstorms if there were no cosmic rays. Ancient man would not have had access to fires and the many thousand deaths each year from lightning strokes would have been avoided.

Cosmic rays originate from deep space and usually consist of very high-velocity atoms that have been stripped of their orbiting electrons. There are also electrons present in space that travel with near-light velocities, but such particles are usually absorbed at very high altitudes in the earth's atmosphere. However, heavy cosmic rays penetrate the atmosphere quite far and often reach the earth's surface. During such an encounter numerous secondary electrons are produced (electron showers) along its track from ionizing collisions with atmospheric molecules. The secondary electrons in turn might ionize a fair amount of molecules themselves before they slow down and attach themselves to atmospheric molecules to form negative ions. The result is that one cosmic particle could be responsible for the creation of as many as one billion ion pairs. Fig. 7 shows the rate of ion production by cosmic rays as a function of altitude.

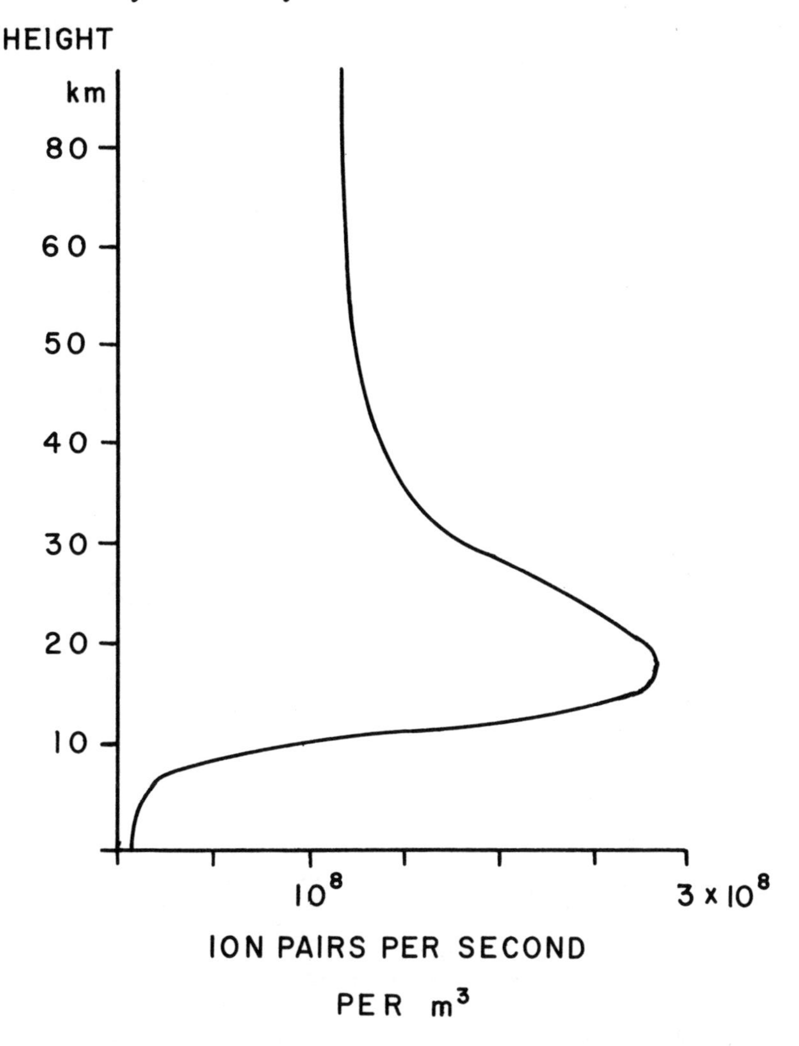


**Fig. 7 Ion production as function of altitude.**

The average production rate at sea level is about ten million ion pairs per cubic metre per second. However, the average ion population at any given time is nearly one hundred times more, which is due to the fact that the average life time of a small ion is about 100 seconds. If all ions in the lower atmosphere were visible they would produce a most dense fog because the average spacing between each ion is roughly 1 mm. If we consider that the ion production rate closely follows the composition of the cosmic radiation at various altitudes (see Fig. 7), we can also expect a marked increase in electric conductivity with altitude. The increase in ion density

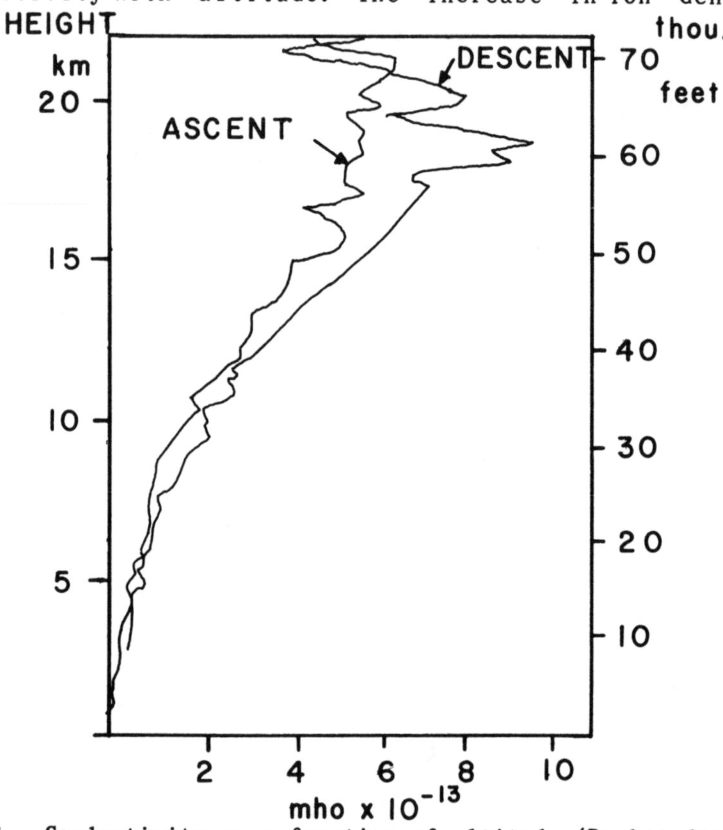


Fig. 8 Conductivity as a function of altitude (Rocket data).

and ion mobility with altitude as air gets thinner has a drastic effect on the electric structure of the atmosphere. Typical conductivity data as a function of altitude are shown in Fig. 8.

The conductivity is the inverse of specific resistance and is usually measured with a Gerdien cylinder (Gerdien 1905). The Gerdien apparatus consists of a cylinder with a coaxial mounted electrode (see Fig. 9). Air is drawn through the

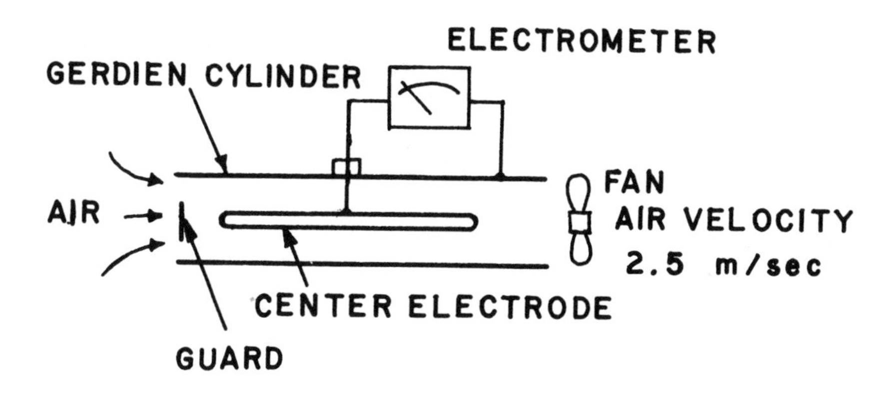


Fig. 9 Gerdien conductivity apparatus.

cylinder at a velocity of a few metres per second by means of a fan. The centre electrode is biased to a few volts via a sensitive electrometer. The amount of current registered on the electrometer relates to the amount of ions per unit volume of air. Care must be taken not to over-bias the inner electrode in order to avoid saturation currents. When properly calibrated, the Gerdien instrument can read both the positive and negative conductivity of air depending on the polarity applied to the centre electrode. A typical ion current plot is shown in Fig. 10 where two identical Gerdien cylinders were tested, the only difference being that one was made of stainless steel and the other of aluminium. Three important features appear from the results shown in Fig. 10. First, the slopes of the two curves are different for negative and positive ion currents. This is due to the difference in ion mobility between negative and positive ions; therefore, the ratio of the slopes equals the ratio of the ion mobilities. Secondly, there is still a negative ion current going to the centre electrode when the electrode is at zero volt bias.

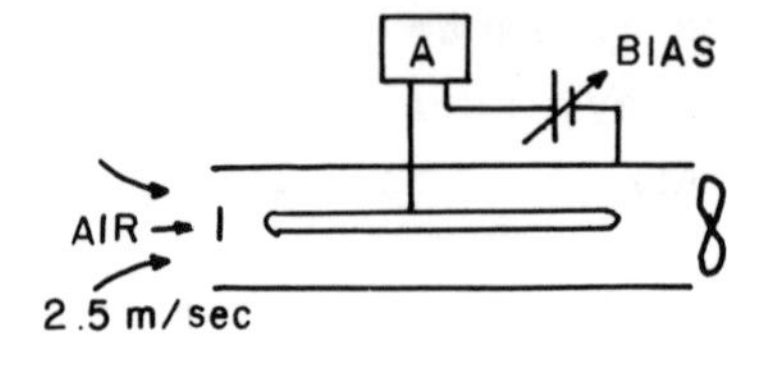


RATIO

| | SLOPES | $I^-/I^+$ at 10V |
|---|---|---|
| SS | 1.25 | 1.36 |
| Al | 1.25 | 1.55 |

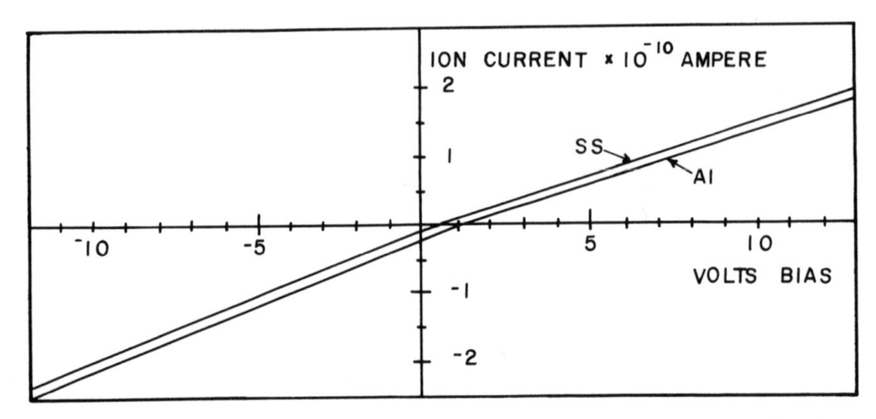


**Fig. 10 Typical ion current readings from a Gerdien cylinder.**

Thirdly, the negative ion current at zero bias is higher for a Gerdien instrument made of aluminium than stainless steel. In order to obtain a zero ion current on the electrometer one needs to bias the stainless steel at 0.4 volts and the aluminium at about 1 volt. These results led to the discovery of electrochemical potentials in the atmosphere (Wåhlin 1971) which appear on material surfaces in contact with ionized air. The reason why stainless steel and aluminium have to be biased at different positive potentials , to achieve zero ion current, is to cancel out the electrochemical or oxidation-reduction potentials which are characteristic of each material and appear when exposed to an ionized environment. The effects of contact potentials are eliminated since both the inner and outer electrodes of the Gerdien instrument are made of the same material. The graphs in Fig. 11 show electrochemical potentials on different materials as a function of positive to negative ion concentration ratio. Electrochemical and contact

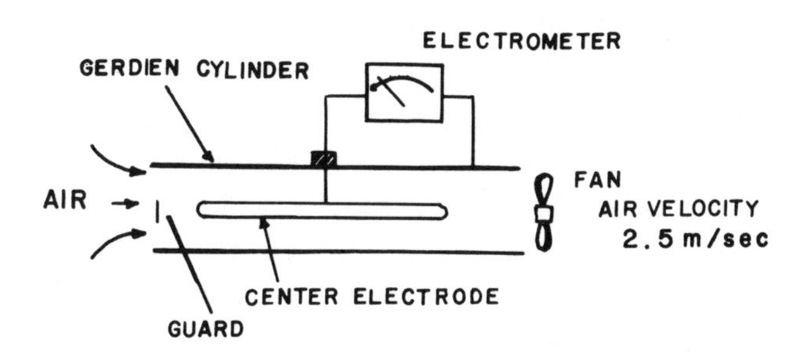


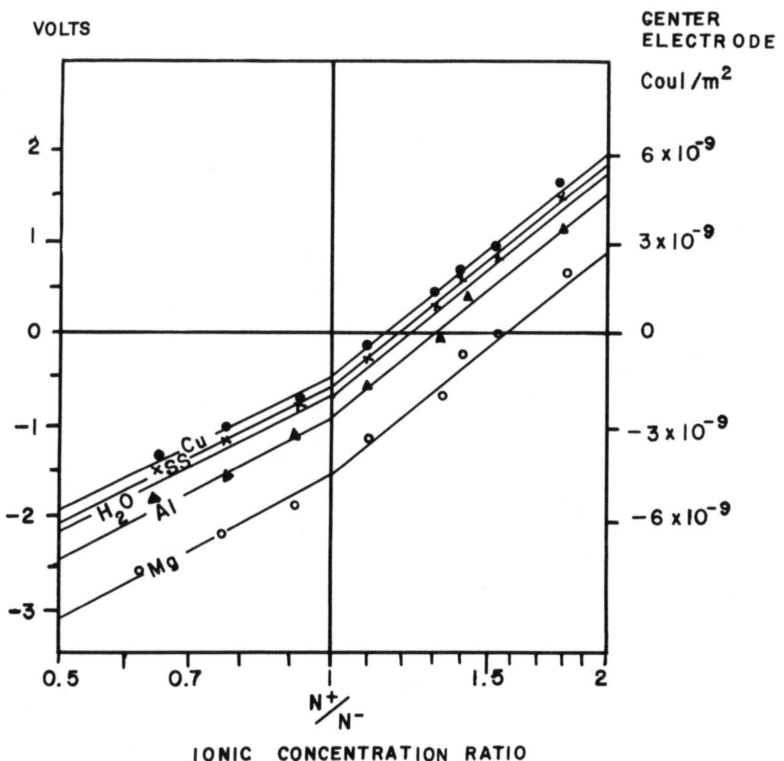


Fig. 11 Electrochemical equilibrium potentials for different materials as a function of ion concentration ratios.

potentials will be discussed further in Chapter 3.

## 2.2 THE FAIRWEATHER ELECTRIC FIELD

The fairweather electric field discovered by Lemonnier and Beccaria (see Chapter 1.2) is almost entirely due to the excess of positive ions over negative ions in the atmosphere. The fairweather field is best understood if we assume that the earth's surface has absorbed a certain number of negative ions from the atmospheric ion pair population. It will create a slight excess of negative charge on the earth's surface with an equal excess of opposite charge in the form of positive ions left behind in the atmosphere. If we imagine that each captured charge on the earth's surface will produce an electric field line which must terminate on a positive excess ion left behind in the atmosphere one obtains a fairly accurate picture of the electric fairweather field in the atmosphere, such as shown in Fig. 12. The excess positive ions are more or less

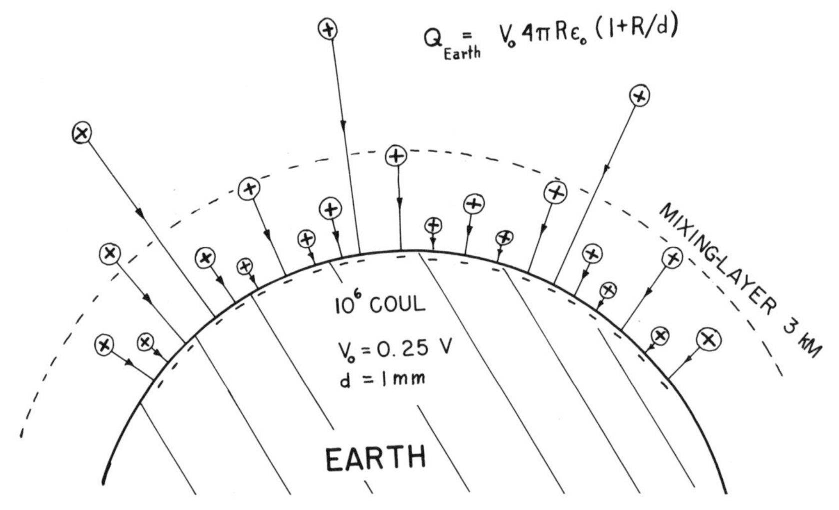


Fig. 12 The fairweather electric field in the atmosphere.

uniformly mixed in the lower 3 km of the atmosphere which, to the meteorologist, is known as the "Austauch" or mixing region. The mixing is produced by convection and eddy-diffusion and the ionic distribution follows the mixing patterns of other constituents in the atmosphere such as radon for example.

Radon is a radioactive gas emitted by the earth's surface and is constantly released into the atmosphere. Fig. 13 shows the vertical distribution profile of excess positive ions compared to that of the radon gas. The positive ion distribution is determined from electric field measurement at varying altitudes and by applying Poisson's equation. The radon profiles are obtained from airborne radioactive counters that detect the daughter products of the decaying radon gas.

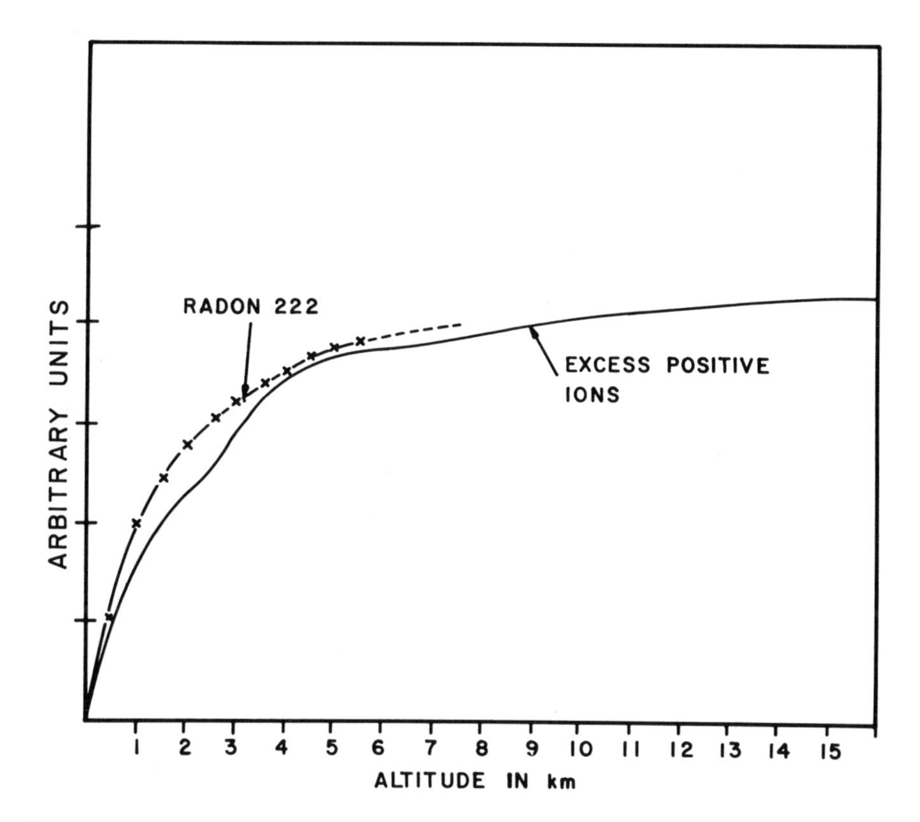


Fig. 13 Relative distribution of radon atoms and excess positive ions as a function of altitude.

The number of field lines per unit surface area produced by the positive charge or space charge above the earth's surface is also a measure of the electric field strength. Obviously the electric field strength reaches a maximum at the earth's surface since it contains the largest number of field lines per unit area. The average field strength at the earth's surface is on average 100 volts per metre and decreases to less than 10 volts per metre at an altitude of 3 km. If one integrates the electric field as a function of altitude one obtains the total potential difference $V$ at different heights. A typical value of $V$ at 3 km is 200 kV with respect to the earth's surface.

The total charge $Q$ on the earth's surface is

$$Q = A E \varepsilon_0 \,, \tag{1}$$

where $A$ is the surface area of the Earth, $E$ the electric field strength at the surface and $\varepsilon_0$ the permittivity of free space ($8.85 \times 10^{-12}$ $Fm^{-1}$). The total energy of the fairweather field is

$$W = \tfrac{1}{2} V Q \,. \tag{2}$$

Fig. 14 shows the total electric energy, charge and potential in the atmosphere as a function of altitude. More than 90% of the energy is confined to an altitude below 3 km which together with the charge distribution curve in Fig. 13 seems to indicate that convection and eddy diffusion play a predominant part in the distribution of the fairweather electric field and that the bulk of its energy is distributed throughout the mixing region by the so called "Austauch Generator" (Kasemir 1950).

One crucial question still remains to be answered. What causes the positive space charge in the atmosphere and how is the opposite negative charge maintained on the earth's surface? As mentioned before there are two schools of thought on this subject, one in which all thunderstorms around the world are believed to charge the earth-atmosphere system (Wilson 1929) and a more recent theory proposed by the author (1973) which considers the electrochemical effect as a charging mechanism where negative atmospheric ions are preferentially captured by the earth's surface leaving a space charge of positive ions behind in the surrounding atmosphere. Both theories might be supported by the evidence of a small systematic diurnal variation in the fairweather field, which is believed to be

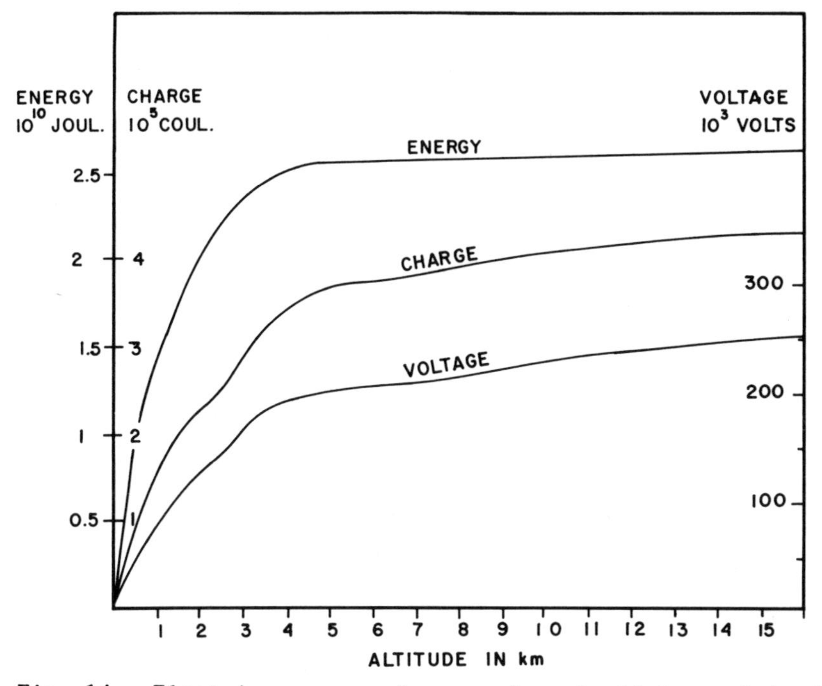


**Fig. 14 Electric energy, charge and potential vs altitude.**

related to the world-wide atmospheric convection activity. The effect was first discovered in Lapland in 1905 by Simpson whose findings were later augmented by Hoffmann (1923) and Mauchly (1923). The effect is illustrated in Fig. 15 where the average variation in the world-wide potential gradient is compared to the estimated world-wide convection activity at different times of day (by GMT). The top graph shows the global variations in the electric field measured at sea in the absence of local disturbances such as pollution, fog, *etc.* The top graph seems to coincide with the lower graph which gives an estimate of the world-wide convection activity produced by the amount of land mass exposed to the heat of the sun during a diurnal period. The steady convection over oceans, however, is thought to smooth out the electric field variations as is evident from the top graph. Before discussing the electrochemical and global thunderstorm circuits as possible generators of the fairweather field, it is necessary to examine the global leakage current and its implications.

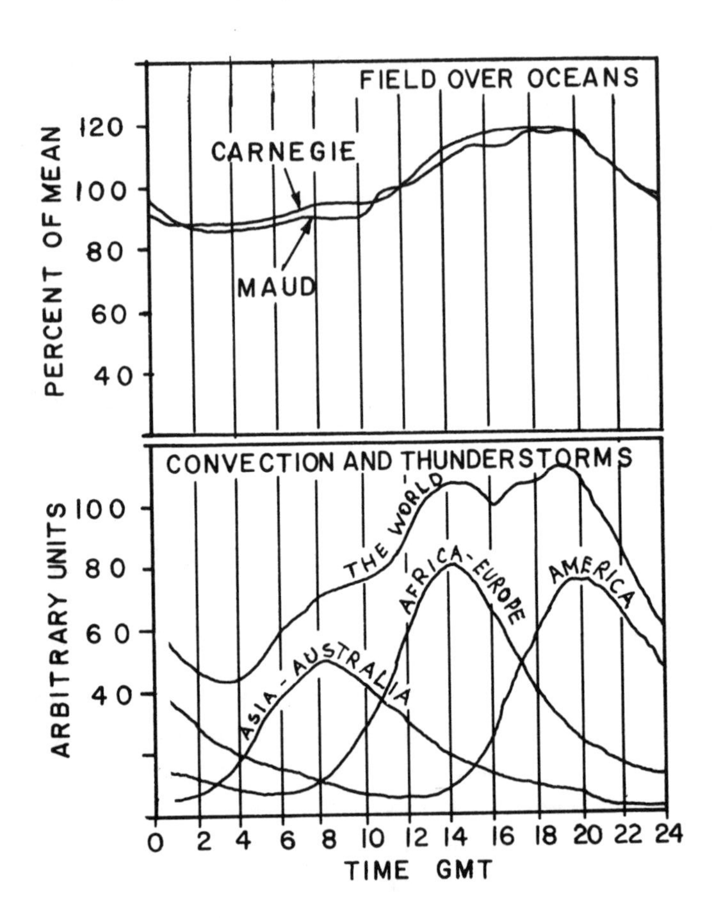


Fig. 15 Diurnal variations in the fairweather field compared to world-wide convection.

**2.3 THE AIR TO EARTH CURRENT**

As already mentioned, the atmosphere is conducting and the earth's electric potential or field must cause a current to flow in the atmosphere. Since there is an excess of positive ions residing in the atmosphere and an opposite negative charge bound on the earth's surface, charge must flow to earth in the form of a positive ion current.

Direct measurements of electric currents in the atmosphere are difficult if not impossible. Therefore, ion current values at different altitudes are almost always computed from conductivity and electric field data by the use of Ohm's law. Direct current measurements can be made, however, at ground level by isolating a portion of the earth's surface and measuring the charge collected over a given time. Several methods can be used (Wilson 1906, 1916, Simpson 1910, Mühleisen 1953 and Kasemir 1951) but in almost all cases the indirect current gives a value often twice as large as the direct method (Lutz 1939, Israel 1954). Whipple (1932) pointed out that the discrepancy in currents can be explained by the fact that there is always convection and eddy diffusion in the atmosphere which will mechanically move charges upwards in the atmosphere thus generating a mechanical or convection current in the opposite direction of the leakage current (the Austauch generator). As later explained, the question whether or not the convection and leakage current on the average are equal is crucial to the electrochemical charging theory and is a problem which has not yet been settled.

From direct current measurements it is possible to estimate the total fairweather current over the whole earth to be nearly 2000 amperes which corresponds to a current density of about $4 \times 10^{-12}$ amperes per square metre. Other charge transfer mechanisms in the atmosphere of importance are point discharges, precipitation currents and lightning discharges.

## 2.4 POINT DISCHARGE CURRENTS

It is difficult to determine the total charge brought to the earth's surface by means of point discharge currents under electrified clouds. Wormell (1930) has made some estimates from the amount of charge brought down by a single point over a period of 4 years. He made a guess that the total point discharge current around the world brings negative charge to the surface at a rate of about 1500 amperes which would supply about 75% of the total fairweather leakage current. Other investigators give slightly lower values for the average point discharge current but not less than 25% of the fairweather current. The source of point discharge currents are the electrified clouds which of course also bring charge to ground by lightning. The point discharge current is, to a certain extent, cancelled by the large amount of positive charge reaching the earth's surface by precipitation.

## 2.5 PRECIPITATION CURRENTS

The electricity of precipitation has played an important role in atmospheric research due to the belief that charging of precipitation particles in some way must relate to whatever charging mechanism is active in clouds. Paradoxically, this is not always true because the final charge on a cloud drop is determined in the space between the cloud base and ground and is usually of opposite sign to the charge of the cloud base where it came from. This peculiar phenomenon is called the mirror-image effect and is demonstrated in Fig. 16 by the two curves which show the change in electric field strength and amount of precipitation charge reaching the earth's surface as a function of time. One can easily see that when the electric field goes negative (negative charge in the cloud base) the precipitation current becomes positive and *vice versa*. As pointed out by Chalmers, a drop must take several minutes to fall from the cloud base to ground. Since the precipitation charge changes with the potential gradient below the cloud, it must mean that the drops also obtain their final charge below the cloud or very near ground.

The electrochemical charging process can possibly explain the mirror-image effect if one assumes that the positive to negative ion concentration ratio near ground is affected by the strong electric field under the cloud. For example, a positive charge on the earth's surface, caused by a strong negative cloud charge above, would attract and remove part of the negative ion population near the surface. The result would be

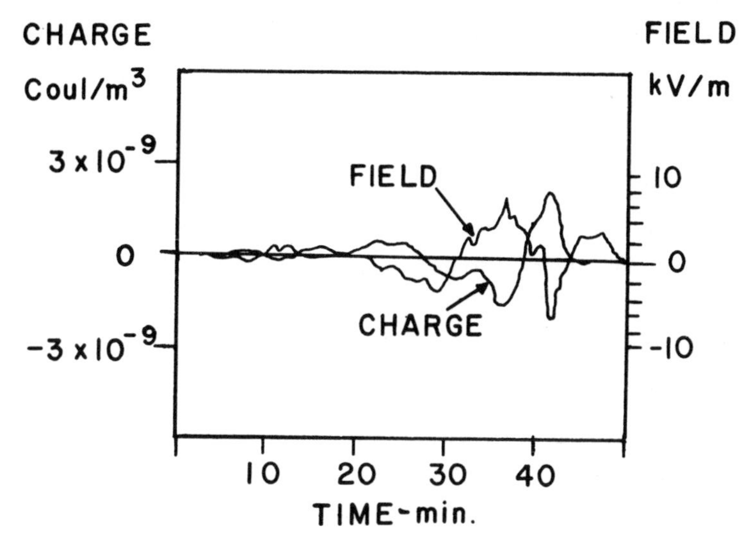


Fig. 16 The mirror-image effect.

a higher than normal positive to negative ion concentration ratio at lower levels. When the positive to negative ion ratio exceeds 1.2 (see Fig. 11) it will produce a positive electrochemical potential on water drops falling through such a region as demonstrated by the Gerdien apparatus experiments in section 2.1. On the other hand, a positive cloud charge above would reverse the effect because drops now fall through an environment containing a higher negative to positive ion concentration ratio which will generate negative electrochemical charges on their surfaces. Other explanations of the mirror-image effect take the Wilson charging mechanism into consideration. This charging mechanism is based on the idea that rain drops become electrically polarized when immersed in an electric field such as under an electrified cloud. A negative cloud charge above will induce a positive charge on the top surface of a drop and the bottom surface will acquire a negative charge induced by the positive charge on the earth's surface. The total net charge on the drop, however, would remain zero. As the drop falls through the ionized region below a cloud it would preferentially sweep up positive ions by its negatively-charged bottom. Calculations, however, show that the Wilson mechanism is too feeble to account for the amounts of charge normally collected by drops (the Wilson charging mechanism is discussed further in Chapter 3). In contrast to rain, precipitation currents carried to ground by snow are usually always negative under potential gradients between ±800V/m (Chalmers 1956). The total precipitation current around the earth is estimated to be about +340 amperes.

## 2.6 LIGHTNING CURRENTS

The charge brought to earth by lightning is estimated to average -340 amperes which would cancel the precipitation current. It must be remembered that a mean current of -340 amperes represents the excess of negative charge over positive charge reaching ground by lightning and that the ratio of negative to positive ground strokes equals about 10:1. The average current in a negative lightning stroke to ground is about 25,000 amperes but the total charge averages only 25 coulomb. Positive ground strokes usually carry as much as 10 times more charge and current than do negative strokes although they are outnumbered by 10:1. The ratio of negative to positive ground strokes seems to vary with global location.

It is believed that about 2,000 thunderstorms are active at one time around the earth which amounts to a total number of 50,000 thunderstorms per day.

### 2.7.1 THE ELECTRIC BUDGET

Where does the energy of nearly 200 million watts come from that is required to maintain the earth-atmosphere electric fairweather field? Are thunderstorms generating the fairweather field by leaking off positive charge from cloud tops to the conducting ionosphere and by bringing negative charge to earth in the form of negative ground strokes and point discharge currents? Or is the electric charge on the earth's surface maintained by the electrochemical charging mechanism in close collaboration with convection and eddy diffusion? These are some of the basic questions that are still in need of answers. Both mechanisms are, in the author's opinion, certainly capable of supplying enough charge and energy to the earth-atmosphere system, but new ideas and more sophisticated measuring techniques are needed in order to find the right answers.

### 2.7.2 THE GLOBAL THUNDERSTORM CIRCUIT

The concept that all thunderstorms around the world generate charge to the earth-ionosphere system was first suggested by C.T.R. Wilson in 1920. The diagram in Fig. 17 shows the global thunderstorm system as interpreted by Wilson. Electric

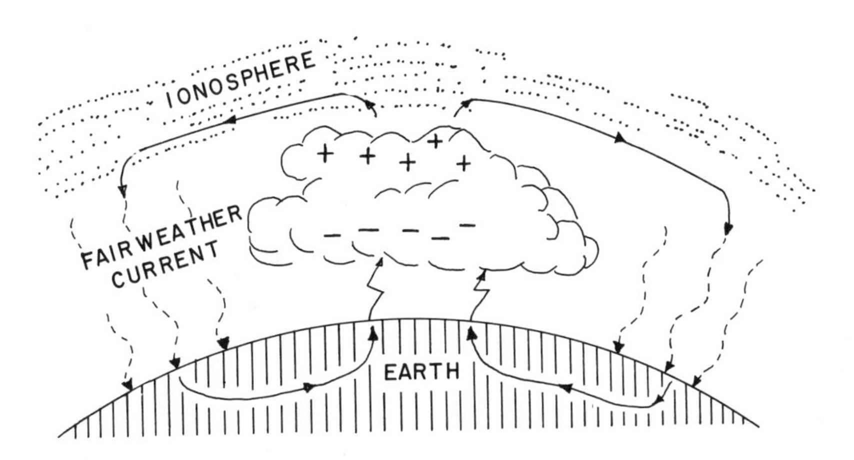


Fig. 17 The global electric circuit by C.T.R. Wilson.

field lines extend from the top of a cloud to the highly conducting upper layers of our atmosphere (50km and higher). Charge is presumed to leak along the field lines from the top of the cloud to the surrounding ionosphere. Note that field lines also go from the top of the cloud to ground thus leaking positive charge to ground. Negative charge is brought to ground mainly by lightning strokes and point discharge currents. The number of field lines between the cloud top and ionosphere compared to the number between cloud top and ground is an unanswered but crucial question which will determine the amount of charge supplied to the global fairweather circuit. For example, if the number of field lines going from the top of the cloud to earth would equal the number of field lines returning from earth to the bottom of the cloud, no current can flow to the ionosphere.

Another problem presents itself when one examines the charge distribution of the fairweather field. Figs. 12 and 14 illustrate that about 90% of the fairweather field and charge is confined within an altitude of 2 km which is far below the conducting ionosphere. The situation is usually explained as follows: consider two conducting surfaces such as the ionosphere and the earth's surface carrying opposite charges at a potential difference of several hundred kilovolts, the earth's surface being negative and the ionosphere positive (see Fig 17). A current driven by thunderstorm generators is flowing in the form of negative ions towards the ionosphere and positive ions towards the earth's surface. Since conductivity and ion mobility increase with altitude, it is believed that negative ions (which flow upwards) will disappear faster on the positive electrode, the ionosphere, than positive ions can disappear on the earth's surface. Positive ions which face an increase in resistance are believed to slow down and congregate in a space charge cloud near the earth's surface. This is believed to produce an excess of positive ions near the earth's surface and could explain the observed positive space charge distribution of the fairweather electric field. There is one problem, however. Plasma physics does not allow ions of one kind to disappear faster on one of the electrodes because it would mean that one of the electrodes or conductors in question would carry more current than the other, which is impossible. Theory and experiments require that an opposite and equal amount of space charge must build up near the other electrode as well (see Papoular 1965). In the case of the fairweather electric field, such a negative space charge near the ionosphere or near tops of thunderstorm generators, has never been found.

Data presented by Imyanitov and Chubarina (1967) provides little support for the closed circuit idea since they show that annual variations in the fairweather field are not in phase with typical thunderstorm activity throughout the world for the same period (see Fig. 18). Furthermore, Kasemir has

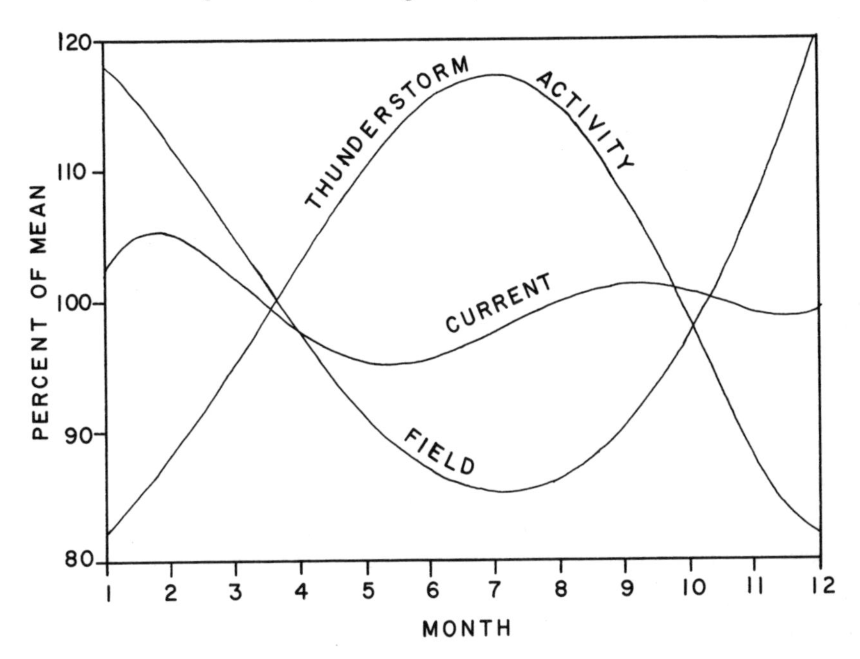


**Fig. 18 Variation in fairweather field and current compared to thunderstorm activity.**

pointed out that the curve showing diurnal variations of the fairweather field in Fig. 15 is much too smooth to fit the world-wide thunderstorm activity data, because recent satelite data show that thunderstorms are mainly active over continental land masses rather than over oceans (Turman, 1978, Turman and Edgar, 1982). The smooth curve in Fig. 15 might be more representative of the world-wide atmospheric convection and follows the field variations of the Austauch generator, a situation which would favour the electrochemical charging theory.

### 2.7.3 THE ELECTROCHEMICAL MECHANISM

The electrochemical charging mechanism considers the earth as an electrode immersed in a weak gaseous electrolyte, the naturally ionized atmosphere. The earth's surface will adsorb negative ions and achieve an electrode potential of about -0.25 volts which will appear at the earth-atmosphere interface in form of an electric double layer. This is analogous to a battery cell where an electrode becomes charged relative to an electrolyte. The thickness of the electric double layer is very small, about 1mm in the normally ionized atmosphere. The double layer can be pictured as containing field lines which connect each captured negative ion on the earth's surface with a positive ion left behind in the atmosphere. As convection and eddy diffusion lift the positive ions to higher elevations the field lines stretch thus increasing the potential with altitude. The result is an electric field build-up which will follow a pattern completely dictated by the mixing and diffusion mechanisms in the atmosphere such as shown in Figs. 12 and 13. The field strength at the earth's surface will equal the electrochemical potential divided by the double layer thickness (0.25volts/1mm = 250 V/m) which is in close agreement with measurements. With a few exceptions (Griffiths and Vonnegut (1975), Moore and Vonnegut (1977) and Willet (1980)) no serious criticism has yet been directed against the electrochemical charging mechanism perhaps because it is a relatively new theory. The above investigators believe that contact potentials rather than electrochemical potentials are responsible for the results reported by the author and that such a charging mechanism is too insignificant to play any major part in atmospheric electricity.

# CHAPTER 3
# Charging Mechanisms

## 3.1 SUMMARY

There are several basic charging processes proposed which in one way or another might contribute to atmospheric electric phenomena. Some of the more important are contact electrification, electrochemical charging, influence charging, diffusion charging and mechanisms involving freezing and splinteringof ice particles. Most of the above charging mechanisms were devised to explain the charging of thunder clouds. Some theories, such as those involving influence charging for example, will not operate in the earlier stages of cloud growth, and others, which involve freezing and ice, cannot be considered dominating since warm tropical thunderstorms exist which do not contain ice. There are numerous other theories which will not be mentioned here and there are even theories which argue that a combination of all charging mechanisms might be at play at one and the same time. The situation is very challenging. The latest theory to be proposed is that of the electrochemical process put forward by the author. The electrochemical charging mechanism has the advantage of being able to explain both thunderstorm charging and fairweather electricity. So far it is the only charging mechanism that can be readily demonstrated by working laboratory models.

Contact electrification involves mechanical contact between solids where electrons from a lower work function material spill over to a higher work function material. Contact potentials are of the order of a few tenths of a volt and might occur in the atmosphere when solid precipitation particles of different temperatures collide or when solid precipitation particles bounce off material surfaces either on the ground or in the atmosphere.

Any process in which charge is captured or transferred by ions is by definition an electrochemical process. Charge transfer by ions can be referred to as oxidation-reduction reactions. An ion which gains an electron is reduced and an ion that loses an electron is oxidized. Electrochemical potentials are encountered in everyday life and can be found in batteries and dry cells for example, and havebeen known to chemists and included in their text books for centuries. One problem, however, is that the language and conventions used by chemists are not exactly tailor-made for physicists who therefore, in the author's opinion, seem to shun the electrochemical effect and often confuse it with contact electrification. An attempt will be made later to explain the difference between contact potentials and electrochemical potentials as seen by a non-chemist.

Influence charging deals with charges that appear on material surfaces which are exposed to an electric field. For example, dust resting on the earth's surface will be negatively charged during normal fairweather conditions since the electric field lines from the positive space charge in the atmosphere above must terminate on the negatively charged earth's surface (the surface in this example being covered with dust). During strong winds the negative charged dust particles can become airborne and form highly electrified dust clouds. Dust and sandstorms are most often found to be negatively charged. The same explanation can be applied to waterfall electricity where the negative charge, induced by the fairweather field on a surface of water, is being carried over the edge of a waterfall. As the electrified water falls over the edge it breaks up into small drops and forms a mist of negative space charge, referred to as the Lenard effect or waterfall electricity, first discovered by Tralles of Bern in 1786. Induction charging in thunder clouds has been considered by several investigators and is based on the idea that cloud drops, which are polarized while subject to the electric fairweather field, preferentially capture negative charge from smaller drops. As large drops fall and collide with smaller drops a transfer of charge is believed to occur where the upper half or negative pole of a smaller drop gives up its charge during the encounter. Negative charge collected in this manner would descend and occupy the lower portion of a cloud while the smaller drops with excess positive charge would remain behind to form an upper positive region.

### 3.2 CONTACT CHARGING

There are many excellent books and papers on contact potentials or Volta potentials, such as Loeb (1958) and Lord Kelvin's famous paper presented at the Bakerian lecture to the Royal Society (1898). But ever since Volta's original observations there has been confusion between contact potentials and electrochemical potentials, a situation which persists even today. For example, Lord Kelvin became very upset when Professor Lodge presented his paper "On the Seat of the Electromotive Force in a Voltaic Cell" (1885) and later in a private letter to Lord Kelvin he expressed his belief that contact potentials are related to the difference in oxidation energies of different materials. Loeb in his book warns, "Volta potentials must never be confused with electrolytic potentials". With due respect to both Kelvin and Loeb (the latter was a colleague of the author's) the author believes that Professor Lodge was also right.

Kelvin's condenser system which is shown in Fig. 19 demonstrates how volta or contact potentials are measured. It consists of two large capacitor plates made of dissimilar

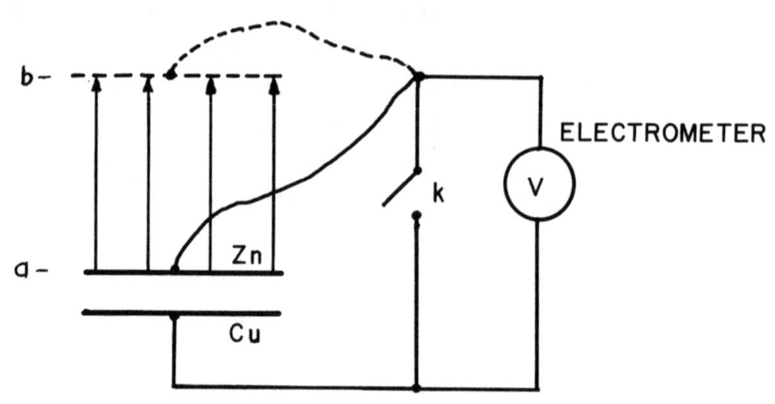


**Fig. 19 Lord Kelvin's condenser system.**

metals (copper and zinc) which are separated by an adjustable air gap. A switch and an electrometer are connected in parallel with the capacitor plates. The plates are connected to an electrometer. At first the switch k is momentarily closed which will allow electrons to flow from the zinc plate to the copper plate because zinc has a lower work function than copper. An electric field with a total potential of $V_c$, which is the difference in work function between the two metals, will appear between the two surfaces. When the spacing between the plates is increased from $a$ to $b$ the field lines will extend and the potential across the electrometer will increase to $V = V_c(a+b)/a$. In the above example the effects of fringing fields and stray capacitance were neglected. The diagram in Fig. 20 illustrates the relationship between work functions and contact potentials. Fig. 20a shows a typical potential well diagram for a metal surface where the conduction and valence electrons are trapped at an energy level which is equal to the work function of the material in question. The potential barrier set up by the work function is represented in Figs. 20b and 20c by small electric cells between the conduction band of the material and its outermost surface. At first, before the switch k has been closed, there is no field between the zinc and copper surfaces. On closing the switch electrons will spill over from the zinc, having the

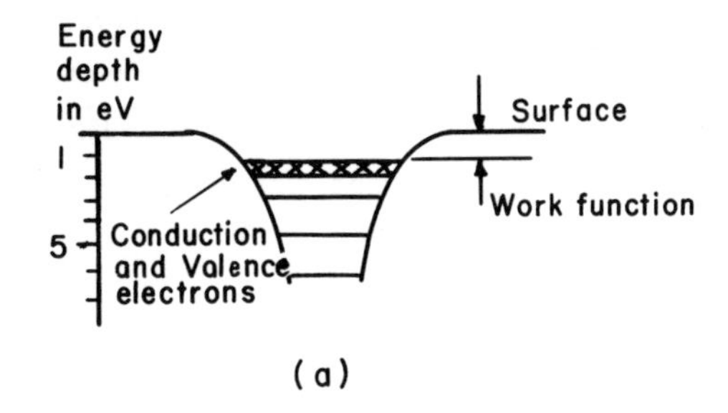


(a)

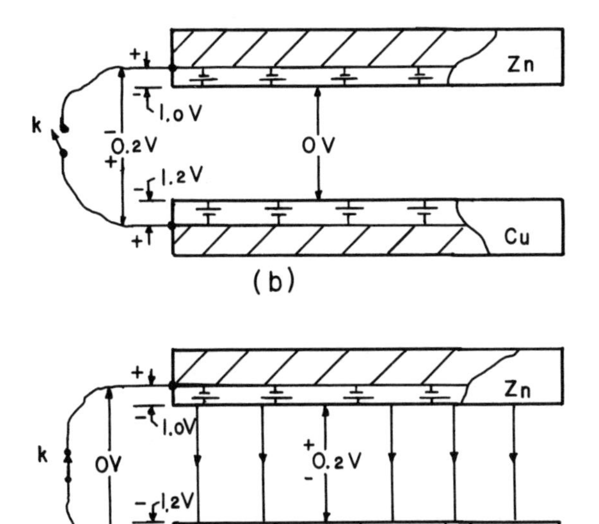


(c)

Fig. 20 Contact potential between zinc and copper.
(a) Potential well diagram of a metal surface.
(b) Electric field before contact is made.
(c) Electric field after contact is made.

lower work function to the higher or deeper potential well of the copper. Bringing the conduction bands of both materials in electric contact will cause the difference in work function potential to appear across the gap of the plates. The amount of charge transfered is $V_c C$, where $C$ is the capacitance of the plates. One characteristic feature of contact charging is that the current ceases as soon as the capacitance $C$ is fully charged; *i.e.* contact potentials are not current driving sources such as galvanic cells, for example. The total amount of energy dissipated in the contact charging process is

$$W = \frac{1}{2}V^2 C . \quad (3)$$

It is interesting to note that in his lecture to the Royal Society, Kelvin mentioned one experiment in which he placed a drop of cold water between the plates in his condenser apparatus and found the electrometer swung toward the opposite direction to that of the contact potential, but with the same magnitude. He attributed this effect to electrolytical conduction and not to contact potential charging. Experiments by Maclean and Goto in Glasgow in 1890 proved that zinc and copper, with fumes from flames passing up between them, gave, when connected to an electrometer, deviations in the same direction, as if cold water had been in place of the flame. Kelvin also mentioned that beside the wonderful agency in fumes from flames, there were reports by other investigators that ultraviolet light and x-rays traversing the gap between the plates caused the same effect as that of cold water. These effects, Kelvin thought, would to some degree fulfill Professor Lodge's idea of some potentially oxidizing process, but, "each one fails wholly or partially to maintain electric force or voltaic potential difference in the space between them". Further communication between Kelvin and Professor Lodge broke down when Lodge, in a letter, said that Kelvin was unrepentant.

Are contact potentials related to the chemical properties of the materials in question or are they purely a physical phenomenon? To clarify this problem one first needs to ask:

where does the energy come from that causes contact potentials to build up? The contact potential which equals the difference in work function between two materials is also equal to the binding energy difference between the electrons in the materials. For example, the electrons in copper are more tightly bound to the atoms in the surface lattice than the electrons in zinc. Therefore, it will require more work to remove an electron from the copper surface than from the zinc surface, thus the name "work function". The chemical binding forces between the metal atoms can be pictured as hooks bonding the atoms together in all directions. However, at the surface there will be dangling bonds because the outermost atoms will have nothing to attach to outside the surface boundary. These dangling bonds make up for the surface energy or work function potentials which are simulated by the electric cells in Fig. 20 b and c. Since the energy of dangling bonds is chemical in nature then any transfer of charge, due to this energy, could technically be classified as an electrochemical process.

What will happen if the space between the plates in Fig 20c is filled with an equal amount of positive and negative ions either in form of a liquid or an ionized gas such as air? Will the negative ions go to the positive plate and the positive ions to the negative plate of the condenser and will there be a continuous current flowing in the circuit as long as the ion supply lasts? The answer is yes. But does the energy that drives the current in the circuit come from the contact potential (potential difference in work function between the materials)? The answer is no. First, the energy available from the contact potential is too minute (see Equation (3)) and cannot sustain a current for very long. It is therefore very clear that the energy driving the current must come from the ions themselves as they interact and recombine with the surfaces. Recombination here means the neutralization of an ion as it loses its charge to a material surface. The electric field between the plates, set up by the contact potential, see Fig. 21a, will draw negative ions to the zinc and positive ions to the copper. Once negative ions enter the zinc, charge is brought to the surface in the form of electrons. These new electrons will immediately spill over

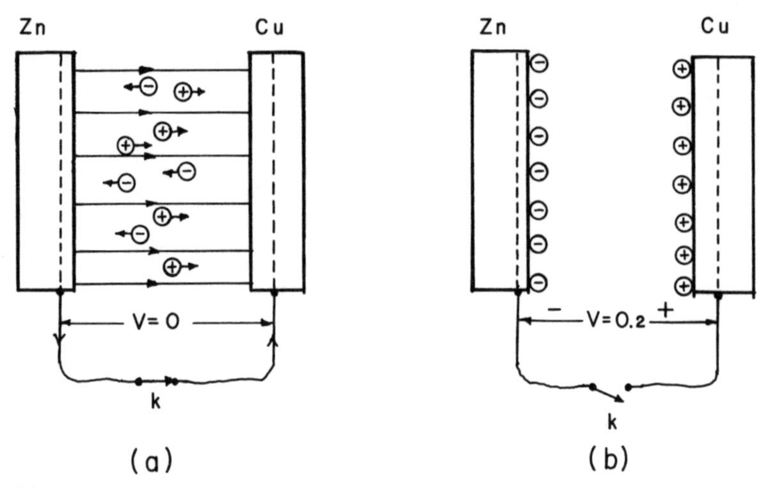


**Fig. 21** Contact potentials in an ionized environment.
(a) Ion current with switch k closed.
(b) Potential build-up with switch k open.

from the zinc to the deeper potential well of the copper work function. However, no constant current can flow unless other electrons are allowed to leave the copper electrode at the same rate new electrons spill over from the zinc; *i.e.* the rate of negative charge brought to the zinc by negative ions must equal the rate of negative charge leaving the copper to combine with the positive ions drawn to the copper plate. What is the source of energy that is capable of lifting the electrons back out of the potential well of the copper work function? It is obviously not the work function or contact potential itself because that would be analogous to lifting oneself by the hair. The energy supplied must come from the ions residing in the spacing between the plates. The ionization or recombination energies of the ions will provide the necessary energy and a current will flow as long as the supply of ions lasts unless switch k in Fig. 21a is opened. If the current is interrupted by the switch, ions will continue to flow and charge the electrodes until a back-emf is built up that cancels the original field of the contact potential across the plates, at which point the flow of ions ceases. A potential equal to the contact potential will now appear across the switch and a situation such as shown in Fig. 21b will be reached.

The presence of ions between two electrodes of dissimilar materials and their ability to generate a steady current is nothing less than a galvanic cell. The primary source of energy is the ionization agency which might be radioactivity or cosmic rays as in the case of our ionized atmosphere. Ions are also produced in liquids where one or both electrode materials might slowly go into solution in the form of ions. The dissolved ions supply the energy that drives galvanic currents through circuits and usually at the cost of the lower work function material itself. A typical example is the flashlight battery or dry cell. Most charging processes involving ions are electrochemical processes.

### 3.3 ELECTROCHEMICAL CHARGING

Electrochemistry has played a major part in both industry and science for the last two hundred years. It might come as a surprise, however, to discover that the function of the familiar dry cell, such as used in portable radios, is not yet perfectly understood. The chemical reactions taking place are believed to be as follows: zinc metal from the outside casing is dissolved by the acidic electrolyte and leaves the container wall as positive metal ions. The charge removed by the positive ions going into the solution will cause a back-emf to build up between the zinc and electrolyte. When the back-emf has reached the same value as the solvation energy, the process stops because the electric field of the back-emf will prevent more ions from going into the solution. The result is an electric half cell with the positive metal ions in the solution forming a tightly bound electric double layer with the negative charged zinc. The other half of the cell is the carbon rod which has the function of supplying electrons to the electrolyte during operation, but the exact chemical reaction involved is not known. The voltage of the cell is determined by the difference in potential between the two half cells. The absolute energy or voltage of each half reaction is not known and no method has yet been devised to measure half cell potentials separately. The problem is how to electrically connect a voltmeter across the double layer without introducing another half cell reaction. Chemists have therefore settled for a compromise method whereby all types of half cells are compared to the voltage produced by a standard cell, the hydrogen half cell. This arbitrary method considers the potential of the hydrogen half cell as equal to zero and the difference in potential between the hydrogen half cell and any other half cell can be found in tables under the heading of Electromotive Force Series. One difficulty is that the potentials listed in the Electromotive Force Series refer to electrodes which are immersed in their own individual solutions containing their own ions, whereas in the dry cell both electrodes are in the same solution; *i.e.* a solution which contains zinc ions but no carbon ions. Another important question is what part does contact potential play in the electrochemical cell? The contact potential is often equal to or very near the potential of the cell itself and contact potentials are hardly ever mentioned in electrochemistry. Are electrochemical and contact potentials so closely related that when Professor Lodge argued with Lord Kelvin 100 years ago, he was right stating that they are of the same nature? It is now known for certain that the valence electrons, which determine the magnitude of contact potentials in metals, also determine the energy involved in electrochemical reactions.

In order to understand the electrochemical charging mechanism, especially as applied to atmospheric electricity, it might be helpful to describe a few laboratory and field experiments which were carried out in an attempt to clarify some of the above problems. Consider the following tests, which can readily be performed in the laboratory, and which are

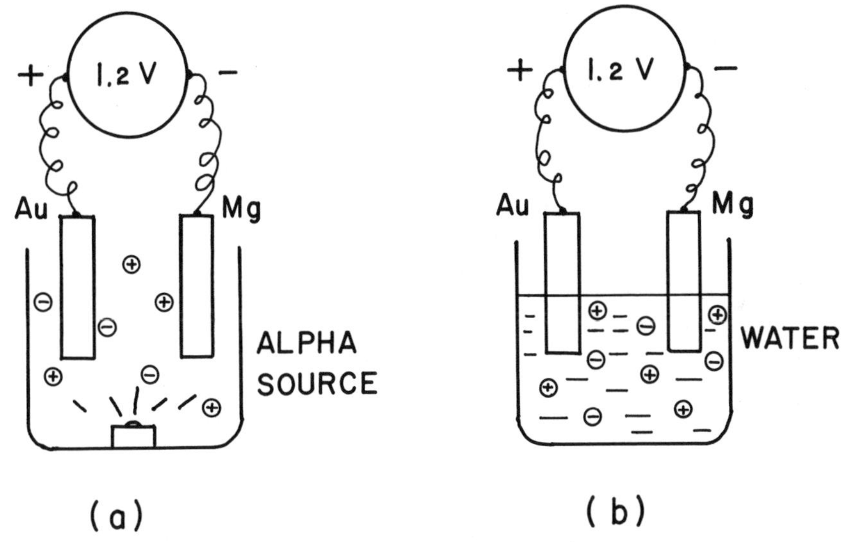


Fig. 22 Experiment demonstrating electrochemical charging.
(a) Galvanic cell with ionized air as electrolyte.
(b) Galvanic cell with water as electrolyte.

shown in Figs 22 a and 22 b:

1. Lower into an empty glass beaker two electrodes of differing materials (*e.g.* gold and magnesium) and connect an electrometer across the two electrodes. When a radioactive source (*e.g.* Po 210, 500 μCu) is placed at the bottom of the beaker, a potential of 1.2 volts will be registered between the electrodes, magnesium being negative with respect to gold.

2. Remove the radioactive source and fill the beaker with sufficient water to partially immerse both electrodes. The electrometer will again register a potential of 1.2 volts and the same polarity.

Both experiments deal with an electrochemical cell, using water for electrolyte in one case and air in the other. The fact that water contains ion pairs makes it an electrically conducting electrolyte which is vital for an electrochemical cell. In the first case, Fig. 22a, the beaker contains air instead of water, but the air is slightly ionized by the radioactive source, and although the number of ion pairs is relatively low, the ionized air is electrically conducting and exhibits the same properties characteristic of an electrolyte.

The average number of ions in the atmosphere is of the order of one billion pairs per cubic metre. The atmosphere is thus very much like an electrolyte. Its ion pairs are produced mainly by the constant bombardment of cosmic rays that reach our atmosphere from solar and stellar sources. Other ions are produced by radiation from radioactive materials in the atmosphere and in the earth's crust (see section 2.1.).

That material surfaces in contact with the ionized atmosphere are subject to electrochemical charging can also be shown by simply probing the atmosphere with electrodes and it can be easily demonstrated that in the absence of external electric fields negative ions rather than positive ions have a tendency to adhere to material surfaces. It can also be observed that when the probes are ventilated by strong winds, more ions are supplied and the negative ion current to the probes increases accordingly. At first, when such measurements were performed it was not at all clear what caused this type of charging although it was believed at the time that work function potentials might play an important part in the charging process. Experiments were therefore set up in the laboratory in an attempt to duplicate the field measurements described. Surfaces of diverse materials were ventilated by artificially ionized air. The results revealed a marked difference in activity among various materials in contact with the ionized air. For example, a small sheet of magnesium metal will absorb negative ions and continue to charge until it reaches a potential of -1.6 volts with respect to its surroundings. Each material tested was found to reach its own characteristic potential. An abridged list of these materials with their potentials appears in Table 1. The potentials in Table 1 agree with the potentials measured by the Gerdien cylinder shown in Fig. 11 section 2.1.

| Material | Work function | Equilibrium Potential Volts |
|---|---|---|
| Magnesium | 3.66 | -1.6 |
| Wet Filter Paper | - | -1.05 |
| Aluminium | 4.24 | -1.0 |
| Cadmium | 4.22 | -1.0 |
| Tantalum | 4.25 | -0.65 |
| Molybdenum | 4.50 | -0.6 |
| Copper | 4.65 | -0.45 |
| Stainless Steel | 4.75 | -0.4 |
| Ice | - | -0.4 |
| Gold | 5.35 | -0.22 |

**Table 1. Work function and electrochemical equilibrium potentials for different materials.**

At first it was disappointing to discover that the potentials in Table 1 were not at all proportional but rather inversely proportional to the work function potentials of the different materials in question. However, Table 1 has a very familiar appearance to the chemist; it resembles a chemical activity series in which the electrical potentials are values related to the oxidation energies of the materials in question. To further prove the electrochemical effect, the same materials were immersed in distilled water, two at a time. Differences of potential between the two materials were then measured and found to be directly proportional to the respective differences between their values in Table 1. From the experiments and the results of Table 1 it became clear that we were dealing with an electrochemical effect and the following conclusions were drawn: negative ions in air are formed from acidic molecules such as $O_2$ and $NO_2$. These molecules are electrophilic (electron seekers) and capture free electrons produced in air by the various ionization processes. The molecular ion, with its valence slightly reduced by a captured electron, still remains acidic and very active as an oxidizer. When the molecular ion oxidizes a surface material, negative charge is transferred by the captured electron to the surface. As more ions reach the surface, a negative back-emf will build up, eventually repelling any incoming ions until no more can reach the surface. Just as in the case of the dry cell an electrical equilibrium potential is achieved. The electrochemical reaction halts when the back-emf has reached a value equal to the oxidation potential, or the energy of the chemical oxidation-reduction reaction involved. Oxidation-reduction reactions can be described as follows: if two elements combine where one wants to share one or more of its electrons with another element which is an electron acceptor, then an oxidation-reduction reaction has taken place. The electron donor is said to be oxidized and the electron acceptor is reduced. It is interesting to note that in Table 1 the elements on top of the activity list are the elements with the lowest work functions. These elements freely share their valence electrons and become easily oxidized because of their lower electron binding energies or lower work functions. Electronegative elements or compounds are the electron seeking oxidizers that want to react with and oxidize material surfaces. The oxidizers or electron seekers appear most commonly as negative ions in the atmosphere and solutions because of their tendency to pick up free electrons in the surrounding environment. When the oxidizer reacts with a material surface it brings a captured electron along. The electron will charge the surface and the chemical reaction can therefore be classified as an electrochemical reaction.

The fact that oxidizers in general appear as negative ions forms the basis for electrochemistry in the atmosphere and in solutions. It is very important to remember, especially in the case of atmospheric electrochemistry, that the process which brings the negative ion to a material surface is the chemical reaction and not the electric field of the ionic charge itself. In an electrochemical reaction the electron simply enjoys a piggy-back ride to the surface and the image force produced by its charge is too feeble and short in range to compete with chemical processes. Also, the strength of the image forces is equal for both positive and negative ions so that no preferential charging of either sign can be expected. Nevertheless, a theory put forward by Phillips and Gunn (1954) considers the difference in mobility between negative and positive ions as possible mechanism for preferential charging. This theory will be discussed in section 3.4.

Numerous experiments in ionized air revealed that all metals and conductors tested proved to be oxidized by negative ions and achieved negative equilibrium potentials. No positive equilibrium potentials were ever encountered. In cases where two electrodes are immersed in an ionized medium, both electrodes become negatively charged relative to the electrolyte, the lower work function material being more negative. The potential between the electrodes equals the difference in oxidation potentials between the materials which also seems to equal their difference in work function except for the reversal of sign. The results were the same for electrodes immersed in ionic solutions where the electrodes did not go into solution. No tests were performed involving solvation energies.

One striking feature of the electrochemical charging process is the formation of the electric double layer. The oxidiation potential and the thickness of the double layer determines the charge density on the oxidized surface. In normally ionized air the double layer thickness is about 1 mm which means that a surface of water, for example, which has an oxidation potential of -0.25 volts, will charge to

$$Q/m^2 = V_0 \varepsilon_0 / d \ , \tag{4}$$

where $d$ is the double layer thickness and $V_0$ the oxidation potential. The value of $d$ was derived from laboratory experiments which proved that the average thickness of the double-layer in air, to the accuracy of the measurements, is equal to the average distance between the ions. It appears that each time a negative ion reaches the surface, it leaves behind a positive ion at a distance equalling the average ion spacing. This distance for normally ionized atmosphere is about one millimetre. In an experiment that followed, isolated spheres (to simulate cloud drops) were ventilated by ionized air for the purpose of studying the charge collection on their surfaces. Negative ions reacting with the surface of a drop or a sphere leave behind positive ions in the surrounding air, thus building up a diffuse space-charge cloud around the sphere. This cloud of positive ions forms an electric double-layer with the charged surface of the sphere. The double-layer, formed by the charged surface of a sphere surrounded by its diffusion cloud of opposite charge, constitutes a spherical capacitor with a capacitance of

$$C = 4\pi R \varepsilon_0 \left(1+\frac{R}{d}\right) \tag{5}$$

$R$ is the radius of the sphere, and $d$ is the mean thickness of the double-layer. The charge on the sphere can then be calculated from the electrostatic expression

$$Q = V_0 C = V_0 4\pi\varepsilon_0 R \left(1+\frac{R}{d}\right) \ , \tag{6}$$

where the double-layer voltage, $V_0$, is also the oxidation-reduction potential of the reaction between the ions and material in question. The sphere with its attached double-layer is, of course, electrically neutral with regard to its surroundings. No drastic charge separation has yet occurred. The fact that the sphere or drop is ventilated (by falling through the atmosphere for example) is of the utmost importance because the flow of air will partially strip away and remove the outer positive layer of the double-layer, a process that not only increases the potential of the sphere but also makes it appear to be negatively charged relative to its general surroundings. For example, removing the outer-charged layer to infinity will increase the potential on the sphere to

$$V = \frac{Q}{C} = V_0 \left(1 + \frac{R}{d}\right) . \quad (7)$$

Common experimental values for charged water drops are shown in Figure 23. These values were obtained from actual

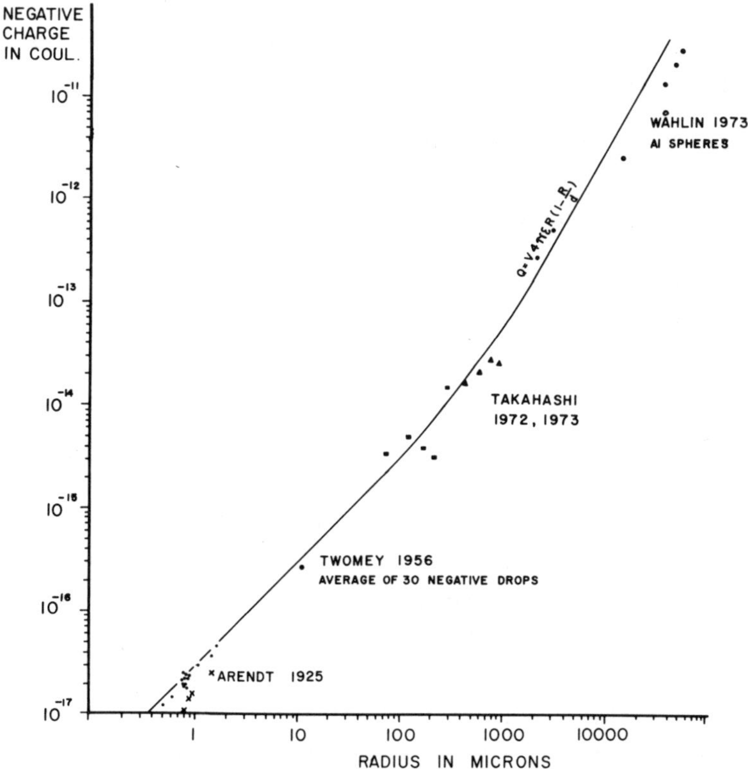


**Fig. 23 Charge-size measurements of rain drops compared to values predicted by the electrochemical charging process.**

measurements in the atmosphere by several investigators and are compared here to the predicted values represented by the solid curve. The curve is constructed from Equation (6), using a double-layer thickness of one millimetre, a distance that equals average ion spacing in the normally ionized atmosphere. An oxidation potential of $V_0$ = -0.26 volts for water was chosen from the experimental results of Chalmers and Pasquill (1937) who, in their laboratory, measured the equilibrium potentials on water drops. The oxidation potential for water in Table 1 is -0.4 volts which is the value obtained for water when ventilated by an equal amount of positive and negative ions. In the normally ionized atmosphere there are more positive than negative ions. A typical ratio is $N^+/N^- = 1.2$ which will lower the equilibrium potential to about one third (see the results in Fig. 11) and which agrees with Chalmers and Pasquill's results.

### 3.4.1 OTHER CHARGING PROCESSES

Except for the well established effects of contact and electrochemical potentials there have been several other charging mechanisms proposed in order to explain atmospheric electric phenomena. Some involve influence charging in combination with colliding drops such as the Elster-Geitel theory, or influence charging and the capture of ions as proposed by Wilson. There are charging mechanisms which consider ice splintering and freezing of water drops to be important in producing charge on precipitation in clouds. The process of evaporation and recondensation of water has intrigued many investigators, including Volta, as a possible source of positive and negative charge in clouds. Takahashi (1973) has recently carried out work along these lines. The validity of the above charging mechanisms is difficult to verify since they lack rigid experimental proof. Recent years have seen many sophisticated computer models which will work if the right parameters are plugged in. This is especially true for influence charging mechanisms involving collision between polarized drops in strong vertical electric fields. Some mathematical computer models are often conceptually difficult. Since mathematics is only a tool of science it is more desirable to have a conceptual theory that is mathematically sound than a mathematical theory that is not conceptually sensible.

### 3.4.2 THE ELSTER-GEITEL PROCESS

The Elster-Geitel (1885) process deals only with the charging of precipitation particles such as drops in clouds. It is based on the assumption that cloud drops, which are polarized in the electric fairweather field, collide with each other and exchange surface charges in a manner that will enhance the fairweather field to a magnitude found in thunderstorms. The proposed charging mechanism is shown in Fig. 24 where a

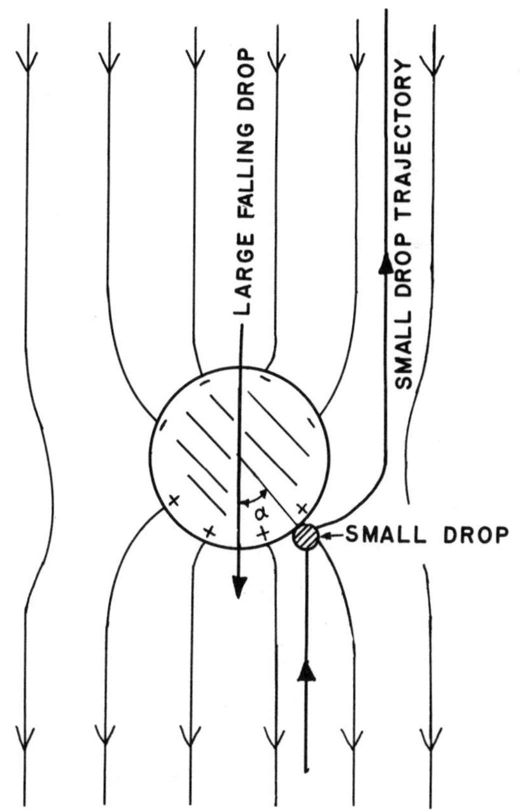


Fig. 24 The Elster-Geitel charging process of cloud drops.

smaller drop which is still light enough to be swept upwards by the updraft winds in the cloud, collides with a much larger and heavier drop falling down through the cloud. Since the large drop is polarized in the electric fairweather field its upper surfaces will attain a small excess of negative charge while the same amount of positive charge will appear on the lower surface. A small drop bouncing off the lower surface will pick up some of the positive charge at the moment of contact, and bring it along to the top of the cloud while the heavier and larger drop, now negative from the loss of positive charge, will continue to fall and bring negative charge to the lower region of the cloud. It can easily be seen that as charge is being separated the electric field strength will increase which in turn intensifies the charging process.

Although sophisticated computer models of the Elster-Geitel theory have been studied (Sartor (1954), Levin (1975) and (1983)) the general feeling is that the charging mechanism is too weak. The theory might possibly work if the initial electric field strength was equal to that found in thunderstorms in which case some other process must prevail (Pathak 1980). Unfortunately there is no laboratory support for the Elster-Geitel theory and some of the unanswered questions are: do small drops really bounce off larger drops at a reasonable rate or do they simply coalesce? How important is the grazing angle of the impact?

### 3.4.3 THE WILSON EFFECT

The Wilson (1929) effect is similar to the Elster-Geitel process with the exception that instead of small drops colliding with larger polarized drops negative ions are believed to be swept up by the larger drops as they fall through the normally ionized air. The positive charge induced on the bottom half of a large drop is believed to preferentially collect negative ions as the drop falls down to lower altitudes thus leaving an excess positive charge behind at higher levels. One serious objection to the Wilson theory is that there are not enough ions produced inside a thunder cloud to account for the amount of charge separated.

### 3.4.4 DIFFUSION CHARGING

Gunn (1957) considered a charging mechanism which is based on the difference in mobility between positive and negative ions. He assumed that since negative ions display a higher mobility in an electric field than do positive ions they must be less massive and therefore also have a greater diffusion coefficient in the absence of electric fields. The expected result is that negative ions will diffuse on to cloud drops at a higher rate than positive ions can. Cloud drops will therefore become more negatively charged and as they fall down to the lower region of the cloud they leave an excess of positive charge behind in the upper region. Experiments with metal spheres ventilated by ionized air were carried out by Phillips and Gunn (1954), see Fig. 25. They confirmed that chrome plated metal spheres exposed to ionized air indeed charge negatively and reach certain equilibrium potentials.

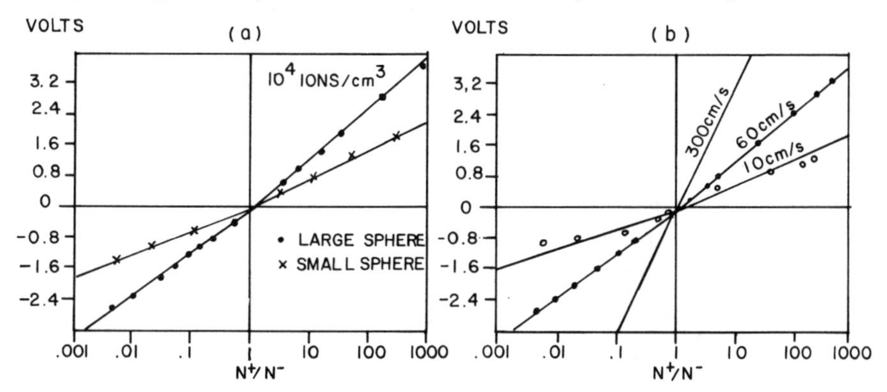


**Fig. 25 Phillip's and Gunn's Experiment with metal spheres.**
**(a) Equilibrium potentials for different dia. spheres.**
**(b) Equilibrium potentials for different air velocities.**

What is interesting about Phillips' and Gunn's experiment is its similarity to the experiments which demonstrate the electrochemical charging mechanism by the use of Gerdien cylinders, see Figs. 10 and 11. In fact, some investigators argue that the equilibrium potentials in Table 1 and the results shown in Figs. 10 and 11 can be explained by the diffusion theory. But this is hard to believe since the diffusion theory does not predict that different materials charge to different equilibrium potentials as shown in Table 1. Also, the energy of the diffusion process relates to the thermal agitation of the ions in their atmospheric environment and is determined by the air temperature. This means that the equilibrium potential, using Boltzmann's constant $k$ and a maximum air temperature of $T = 20^{\circ}C$, cannot exceed $\frac{3}{2}\, k\, T \times 0.25$ or 0.008 volts where 0.25 represents the 25 percent higher mobility of negative ions over positive ions. This is only 3% of the values shown in Table 1 and the graphs of Phillips and Gunn (Fig.25).

### 3.4.5 FREEZING POTENTIALS

Although it is known that warm clouds (clouds that do not contain ice or frozen precipitation) can charge to considerable potentials and occasionally produce lightning, there is still a great deal of attention devoted to charging processes that might involve icing. One reason for this is the observation by Simpson and Scrase (1937) who noticed that separation of charge in thunderclouds seems to occur in regions where the temperature is below freezing. They thought that colliding ice particles might charge negatively and leave positive charge behind in the air in the form of positive ions. The idea was further developed by Reynolds (1954). Workman and Reynolds (1950, 1953) have also discussed a theory in which glazing processes were assumed to produce charge. Many theories have been put forward which involve freezing of water but, again, there are unfortunately no convincing laboratory experiments.

# CHAPTER 4
# The Thundercloud

## 4.1 THERMAL DYNAMICS

On a warm and sunny day heat is absorbed by the earth's surface causing both water vapour and hot air to rise to higher atmospheric levels. The warmer the air the more water vapour it can hold before saturation. The secret of a thunderstorm and its devastating power lies in the amount of water vapour present. The latent energy of water vapour is the key factor in thunderstorm formation and electrification. At the beach most of us have experienced the sensation of cold after getting out of the water. The explanation is that water takes away heat from the body surface at a rate of about 540 calories or 2250 joules per gramme when it evaporates. When water condenses to form drops, the same amount of energy, in the form of heat, is released. When a cloud is formed by warm humid air reaching higher and cooler levels, condensation will create small drops between 5 - 10 microns dia. During condensation latent heat is released, warming the surrounding air which will rise further (the hot air balloon effect) to considerable heights, pulling more humid air up from below to take its place. A chain reaction starts when humid air is fed in from below pumping more and more energy into the cloud in the form of heat and convection. A cloud might grow up to 60,000 feet or 20 km in height in very narrow columns. Updraft velocities can easily reach 30 m/s. Each column is called a cell and a thundercloud usually consists of several cells. Each cell has an average life time of 30 minutes. When a cell reaches a maximum altitude it is in a so-called mature stage, at which time the cloud top usually flattens out to form the typical anvil shape often associated with thunder heads. The flat top is believed to occur when a cloud reaches the stratosphere or the boundary in the atmosphere where air temperatures begin to rise with altitude which will cause the cloud to evaporate again. In the growing stage of a cloud, drops increase in size as they reach higher levels and eventually become heavy enough to overcome the updraft velocities of the convection. At this point rain or precipitation starts to fall down and the cell has reached its dissipating stage. At greater latitudes, thundercloud precipitation without exception always reaches temperatures below freezing. Hail and sleet are formed which often fall down and reach ground before melting. The cold precipitation will cool the air in the cell which becomes heavy and begins to move downward as well. The cold heavy air might cause considerable down drafts which, when reaching ground, supply the outward rush of cool air often felt as a relief to most of us after a hot day. When precipitation and down drafts occur the cell is said to be in its dissipating state and although some lightning activity is still going on its life is nearly over. New cells are formed adjacent to the old ones and often while observing their growth one can see small bubbles or turrets developing above and around each cell. The thunder cloud described is a typical heat cloud or isolated storm. Also common are lines of thunderstorms or squall lines. Lines of thunderstorms are formed when a cold front wedges in under the warm humid air mass along a warm front and squall lines can extend for several hundred miles. The energy source is the same, namely condensation of water and release of latent energy. Thunderstorms along a squall line often seem to lean over on their sides because of the heavier and cooler air wedging in underneath the warm front. Although there is considerable entrainment of air through the sides of such storms, the main energy is still being supplied by the updraft winds caused by condensation and convection.

### 4.2 THE ENERGY OF THUNDERSTORMS

The energy of a thunderstorm is determined by the amount of water vapour present. Typically, a single cell contains $8 \times 10^{8}$ kg of water (Israel 1973) of which about 60% is converted into heat and precipitation. The total energy dissipated by a cell is therefore $1.4 \times 10^{14}$ joules per cell. If the average life time of a cell is 30 minutes then the average power per cell equals $7.8 \times 10^{10}$ watts. Since it is estimated that about 2,000 thunderstorms are active at the same time around the earth then the total energy dissipated must equal about $1.6 \times 10^{14}$ watts which is about 0.1% of the energy reaching the earth's surface from the sun.

### 4.3 THE ELECTRICAL ENERGY IN THUNDERSTORMS

The electric energy of a thunderstorm can be determined in several ways. Schonland (1950, 1953), for example, estimated the potential difference involved in a typical lightning flash to be about $10^{8}$ and $10^{9}$ volts while the average charge transferred is about 20 coul. The total amount of free charge separated in a thundercloud is about 1400 coul according to Wormell (1953). From the above information we can determine the total electric energy involved to be of the order of $\frac{1}{2}VQ = 7 \times 10^{11}$ joules. A more realistic method presented by Israel (1973) yields a larger value of $1.7 \times 10^{13}$ joules. In Israel's approach the average thunderstorm cell is compared to an electric circuit diagram (Kasemir 1965) such as shown in Fig. 26. According to Wait (1950), on the other hand, a thunderstorm cell generates on the average 2.5 amperes of current and from Ohm's law one then finds that the total power output of a cell is $R I^{2} = 9.4 \times 10^{9}$ watts, which when integrated over the duration of a cell (30 min) amounts to

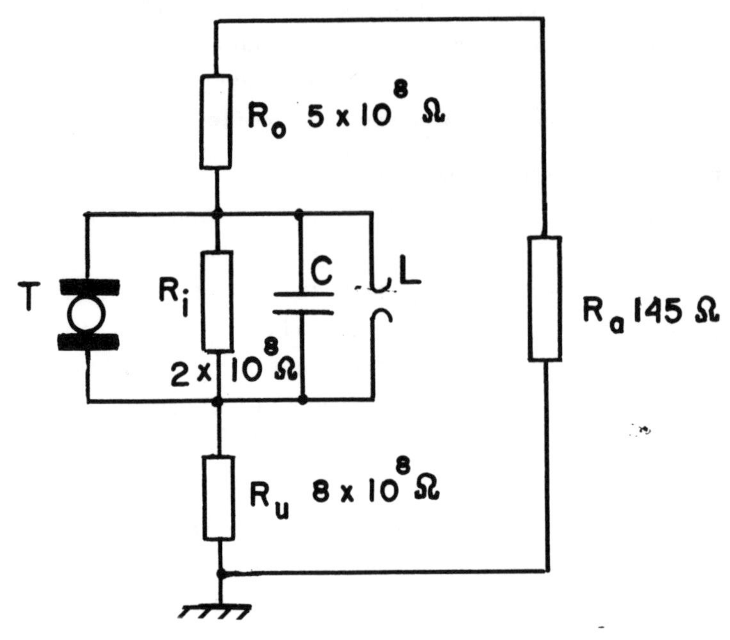


Fig. 26 Electric circuit diagram of a thunderstorm (Kasemir).

$1.7 \times 10^{13}$ joules.

There is at least one serious objection to the use of Ohm's law in determining the electric power of a thunderstorm. There is no doubt that the internal resistance of a thundercloud, such as represented in Fig. 26, is an accurate measurement since it has been determined *in situ* by conductivity measurements. The problem is that the conductivity or resistance depends on the number of ions available in the cloud. Ions are the charge carriers in the cloud and solely responsible for the flow of electric current. Considering that the normal ion production in the atmosphere is about ten ion pairs or $1.6 \times 10^{-18}$ coul of charge per cubic centimetre per second and that the average thunder cloud cell measures fifty cubic kilometres in volume, then the total charge production within the cell is only 0.08 coul per second which translates to a meagre saturation current of 80 mA. Ohm's law will not hold once a state of saturation has been reached. An ohmic current of 2.5 amperes used by Israel might, therefore, not be a representative value. If the charging current in thunderstorms is non-ohmic the only other alternative is a mechanical transfer of charge by means of convection and updraft currents or by the process of gravitation in which charge is transported by falling precipitation particles. The author has reason to believe that when Israel calculated the average electric power involved in charging of clouds he not only considered the ohmic current but also took the mechanical currents into account.

An interesting thought is that if one could harness a lightning bolt there would be 100 kW hours of energy available, enough to keep a 100 watt light bulb burning for a month and a half. There are nearly 50,000 thunderstorms active around the world per day, and if each storm produced 100 lightning bolts, there would be close to 20,000 megawatts, which is only enough power to satisfy the needs of a large city such as New York.

**4.4 THE LIGHTNING DISCHARGE**

There are two types of lightning discharges, namely cloud to ground lightning and intracloud lightning or cloud to cloud discharges. As the reader recalls from section 1, Benjamin Franklin observed that thunderclouds tend to form gigantic electric dipoles in the atmosphere where the lower portion of the cloud is predominantly negative and the top positively charged. Occasionally a small pocket of positive charge can be found in the lower cloud section during the cloud's dissipating stage in the vicinity of the precipitation shaft. There are about three times as many intracloud discharges as cloud to ground flashes, a ratio which seems to vary somewhat with geographical location. Normally negative charge is brought to ground by lightning that originates from the lower section of a cloud. However, as more and more negative charge is drained by successive lightning bolts to ground, the more the positive charge in the upper half of the dipole becomes dominant. The imbalance of the dipole will eventually cause a positive discharge to appear from the upper region of the cloud to ground. The ratio of positive to negative ground strokes is about 1 to 10 but a positive discharge from the upper region of the cloud is as a rule ten times more powerful and carries ten times more charge.

A lightning flash to ground might consist of several ground strokes which appear in such a rapid sequence not distinguishable to the naked eye but easily observable with the aid of a Boys camera (see Fig. 6). In fact, pictures taken with a Boys camera reveal quite a complicated discharge mechanism of lightning. It appears that lightning is triggered by faint pilot streamers (Schonland, 1938) which provide the initial ionized path from cloud to ground by weak corona discharges.

The pilot streamer is usually followed by a stepped leader which can be described as small current surges catching up with the pilot streamer in small steps averaging 50 metres in length. As the pilot streamer works its way down to ground, followed by the stepped leader, charge is brought down at a rate of 600 to 2,600 amperes (Hodges, 1954). As soon as the pilot streamer and stepped leader reach ground a conducting path between cloud and earth is established and the main stroke starts. The main stroke will transfer charge between the cloud and earth at a rate of 20,000 or sometimes 400,000 amperes. One peculiar behaviour of the main stroke is that it starts from ground and travels upwards which means that large amounts of charge are transferred from the earth's surface to the cloud. This is why it is often said that lightning travels up to the cloud rather than down. Although the main stroke only lasts for a thousandth of a second, it is visible to the naked eye and the sound or acoustic shock wave that follows is unmistakably that of a lightning bolt. The main stroke often breaks up into several subsequent strokes, usually three or four per lightning flash. Lightning flashes having as many as 42 subsequent strokes have been reported. There are usually no pilot streamers or stepped leaders present between ensuing strokes in a multi-stroke flash since an ionized path has already been established after the first stroke. The stepped leader is replaced by a so-called dart leader which works its way down from the cloud in one single jump.

One of the most challenging problems in lightning research is the mystery of how lightning bolts drain charge from the thunder cloud. How does the lightning bolt connect up to the myriads of charged drops involved, and how can this be accomplished in a small fraction of a second? This is a question that has not yet been answered satisfactorily. One of the best solutionsso far, in the author's opinion, is offered by Kasemir (1950b) who believes that the mechanism involved is

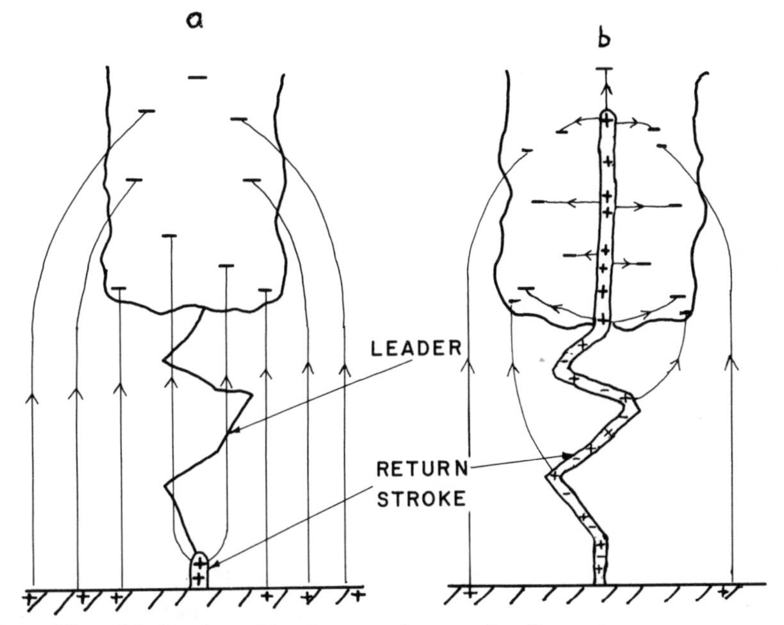


Fig. 27. Lightning discharge theory by Kasemir.
(a) Start of return stroke.
(b) Charge transfer by induction.

that of induction. For example, consider a negative charge centre in the lower portion of a cloud (see Fig. 27a.) from which a leader stroke has just reached ground. Not much charge is brought down because the negative charge in the cloud is bound on small drops or precipitation particles. On the other hand the earth's surface, which is a good conductor, can easily supply positive charge up to the cloud through the ionized channel provided by the stepped leader. When the positive charge in the return stroke reaches the cloud it will penetrate the charge centre like a giant lightning rod which causes the cloud potential to drop drastically since many field lines now are much shorter (see Fig. 27b). In this way much of the electric energy is drained without contact or charge being removed from the drops in the cloud. Another way to look at Kasemir's lightning mechanism is to compare the electric potential and capacitance of the cloud before and after a lightning bolt has penetrated the cloud. A 25 coul charge centre measuring 1 km in diameter and located several kilometres above ground has a capacitance of roughly

$$C = 4\pi\varepsilon_0 R \approx 6 \times 10^{-8} \text{ Farad} \quad , \tag{8}$$

which yields a potential of

$$V = \frac{Q}{C} \approx 4 \times 10^{8} \text{ volts.} \tag{9}$$

If one assumes that the charge centre is cylindrically shaped and extends 1km in height then the capacity between the charge centre and the ionized channel of the lightning bolt can be estimated from

$$C = \frac{2\pi\varepsilon_0 l}{\ln(b/a)} \approx 9 \times 10^{-9} \text{ Farad,} \tag{10}$$

where $a$ is the radius of the lightning channel and $b$ the average distance of the charges in the charge centre with respect to the lightning channel. The radius of the lightning channel was taken as 1.5 m. The result is that the capacity is lowered by about 15 percent which corresponds to a charge transfer of approximately 3.5 coulombs. Since there are on average three or four main strokes per lightning flash, then the total charge transferred by a lightning flash can easily be between 10 to 15 coulombs. In fact, Kasemir's capacitor model does not only explain why the bulk of charge is moving upwards, from ground to cloud under the influence of induction, it also presents us with a $R\ C$ circuit which might explain the occurrence of multiple strokes. The time constant of the circuit (average 200 μs) is the product of (9) and (10) divided by the average discharge current of 20,000 amperes. Once the capacitor is charged the lightning current will cease and the time elapsed is of course determined by the time constant. After an average length of 0.07 s (Bruce and Golde, 1942) the next stroke will follow but this time triggered by a dart leader. The ensuing stroke will form a branch which will penetrate the cloud in a different direction. More strokes might follow until the negative potential of the entire region has been drastically lowered. The induced charge transferred in this manner must slowly leak off to the individual drops by means of corona and quiet discharges that will produce numerous ion pairs inside the cloud.

In an intracloud discharge a lightning bolt starts between two charge centres where the electric field is high and where electric breakdown of air first occurs. After a small parcel of air becomes conducting from the electric break-down of the air, it will grow rapidly in length under the influence of induction. As the lightning rod type discharge grows and becomes more and more polarized, positive charge will rush towards the end facing the negative charge centre while negative charge collects at the end nearest the positive charge centre. Once the lightning bolt penetrates both charge centres, enough charge is transferred by the polarization effect to account for the observed electric field changes.

One important implication of Kasemir's discharge mechanism is that the electric breakdown of air, and formation of small ionized air parcels in the strong field region between two charge centres, can trigger lightning bolts. This means that aircraft flying in the high field region between two charge centres can become electrically polarized and trigger lightning as well. The risk of damage is therefore much greater than previously thought when it was believed that lightning would only strike aircraft in a random fashion. Pilots usually refer to two kinds of lightning strokes; the "static discharge" when lightning seems to originate from the aircraft when it is trying to rid itself of excess charge or the direct hit when lightning is believed to come from a nearby charged cloud centre. The "static discharge" is controversial in that the charge on an aircraft, as pointed out by many scientists, is by far too small (about 0.1 coul) to create a lightning bolt. Kasemir's induction mechanism, however, explains that the strongly polarized electric field on the aircraft, which causes corona discharges and electric break down of air, will supply all the charge necessary for the lightning bolt. For example, although the net charge on the aircraft is small, the amount of ionic charge produced in the surrounding air by corona is enough to feed a lightning bolt by the inductive mechanism. One cubic centimetre of completely ionized air contains as much as 5 coulomb of ionic charge.

The size of a lightning bolt or the diameter of the lightning channel is not well known. A rough idea of the channel diameter can be obtained if one uses Gauss's law to determine the maximum radius of electric break down of air due to the strong field and corona produced by the charged lightning channel. For example, an average lightning bolt which carries a current of $I = 25000$ amperes and propagates at a typical velocity of $v = 10^8$ m/s will contain a charge of

$$\lambda = I/v = 2.5 \times 10^{-4} \text{ coul/m} , \qquad (11)$$

which by the use of Gauss's law equals a channel radius of

$$R = \lambda/(2\pi\varepsilon_0 E) \ , \quad (12)$$

where $E = 3 \times 10^6$ v/m is the breakdown field of air. It is believed that the bulk of the lightning current is confined within a small core having a cross section of about 0.15 m in diameter. The lightning channel can be pictured as narrow current core surrounded by a large ionized envelope.

## 4.5 PROTECTION AGAINST LIGHTNING

Modern lightning research has proved that the function of a lightning rod is not necessarily to attract lightning and lead it to ground in order to dissipate its power. Careful studies show that large amount of corona discharges occur at the point of a lightning rod before lightning strikes. The corona discharges produce numerous ions which often develop a protective dome around structures fitted with lightning rods. A large conductive dome or sphere has a smoothing effect over protruding surfaces and will lower the electric field strength in its vicinity. Experiments by Vonnegut and Moore have shown that a conducting wire supported between two mountain tops did not get struck by lightning as often as the mountain sides. In fact, the lightning seemed to avoid the wire which can be explained by the above effect that corona must produce a large diameter ion cloud around wire. This has the effect of lowering the field strength in its vicinity.

The lightning rod is an important device for protecting houses and other structures from lightning. This is accomplished in two ways. First, corona from the sharp tip of a lightning rod creates a protective ion cloud above the structure and secondly, in the event of a direct hit by lightning, the lightning rod will force the electric power to dissipate into ground and prevent damage to the building. When protecting a house or structure from lightning damage, some fundamental rules should be followed. Basically, there are two types of protection, internal and external. External protection concerns the safety of the roof and sidings of a house while interior protection deals with sparks between metal object inside the house including appliances, radios and telephone lines. An effective exterior lightning protection can be described as follows:

1. Lightning rods and guard wires must be symmetrically placed over the roof to ensure uniform distribution of the discharge current.
2. Lightning rods and guard wires should be symmetrically attached to vertical conductors which distribute the discharge current to ground.
3. A good ground connection, far away from under-ground cables and pipes, must be established in order to let the energy dissipate into the earth's surface as effectively as possible.
4. No point on a roof should be further than 10 m (30 feet) away from a lightning rod or guard wire.
5. All large metal objects on the roof should be electrically connected to the system.
6. Any large protruding non-conducting structures or objects on the roof should be fitted with an individual lightning rod.
7. A building should have at least two vertical conductors to ground.
8. Sharp bends or curves in the conductors should be avoided since they might cause unnecessary induction and build-up of sparks.
9. Large metal objects on the outside walls such as balconies and fire escapes must be connected to the system.

The recommended cross sectional area of guard wires and vertical conductors should be at least 25 $mm^2$ (0.04 $in^2$) and twice that area for the buried wires that serve as ground connections. Fig. 28 shows several different ways of protecting houses and similar structures. Fig. 29 demonstrates a simple lightning protection for sail boats.

Thousands of people around the world are killed by lightning every year. Most fatalities occur in open fields and usually when lightning strikes within a range of 50 to 100 metres of a victim. The danger is that when lightning strikes ground a tremendous amount of current radiates out in all directions from the point of impact. Since the current travels on the

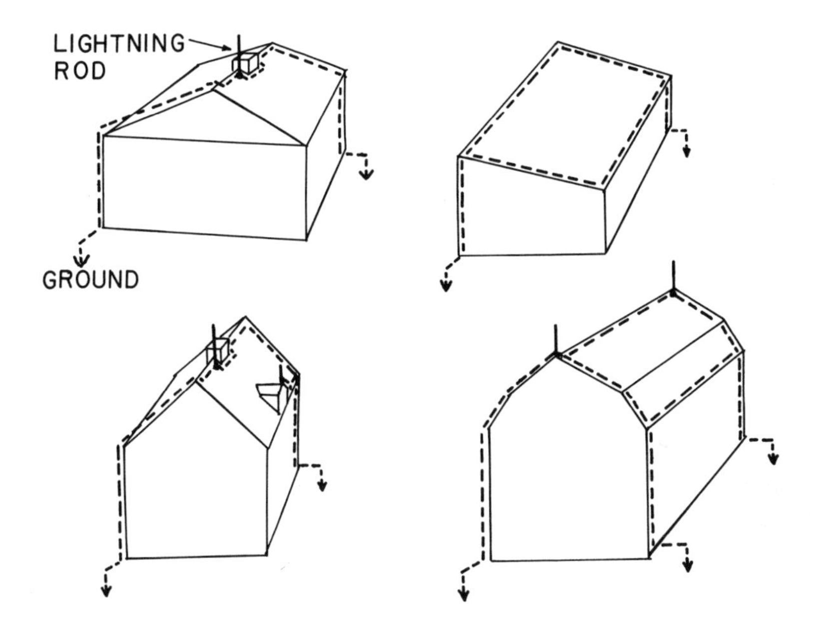


**Fig. 28 Lightning protection for houses.**

surface, which offers a certain electric resistance, a voltage drop will develop which can easily reach 1,000 volts per metre 100 metres from the point of impact. A person walking might bridge a lethal voltage in one step or could easily be thrown off his feet by muscular contractions from the electric shock. It has often been said that it is safer to run than to walk across a field during an overhead lightning storm. This might not be far from the truth since only one foot would be in

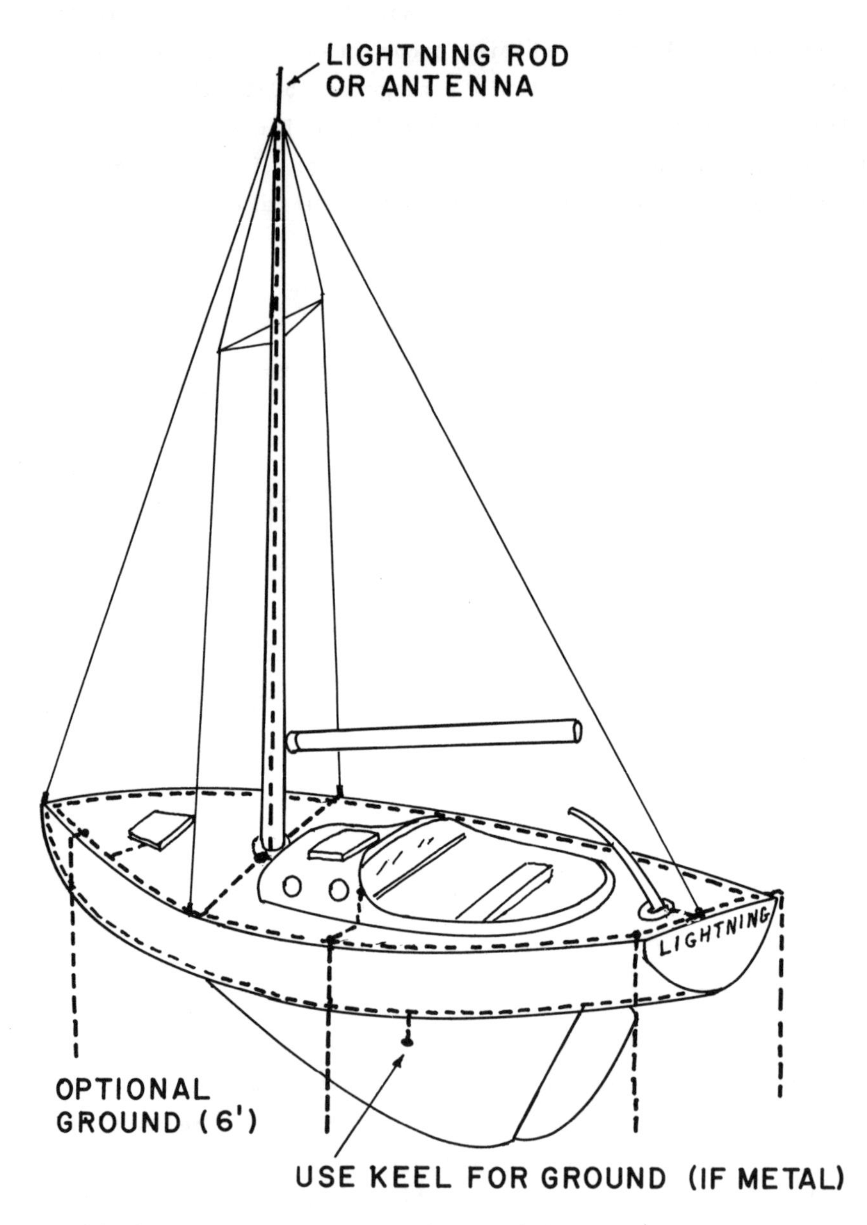


Fig. 29 Lightning protection for sail boats.

contact with ground in case lightning strikes. Lightning often strikes trees, consequently one should never take shelter under tall trees, near power lines, etc.

## 4.6 THE ELECTRIC CHARGING OF CLOUDS

Numerous observations and measurements have shown that a thundercloud theory must satisfy certain conditions:

1. It must explain a positive charge in the upper region of the cloud and a negative charge in its lower portion and during the later stage in the cloud's life cycle it must justify the appearance of a small pocket of a positive charge near its base.
2. The theory must clarify why it takes approximately 20 min. to generate enough charge for the first lightning flash and only a few second for the ensuing flashes.
3. The process must be capable of separating a charge of at least 1,000 coulomb per thunder cloud during an average life time of 30 minutes.

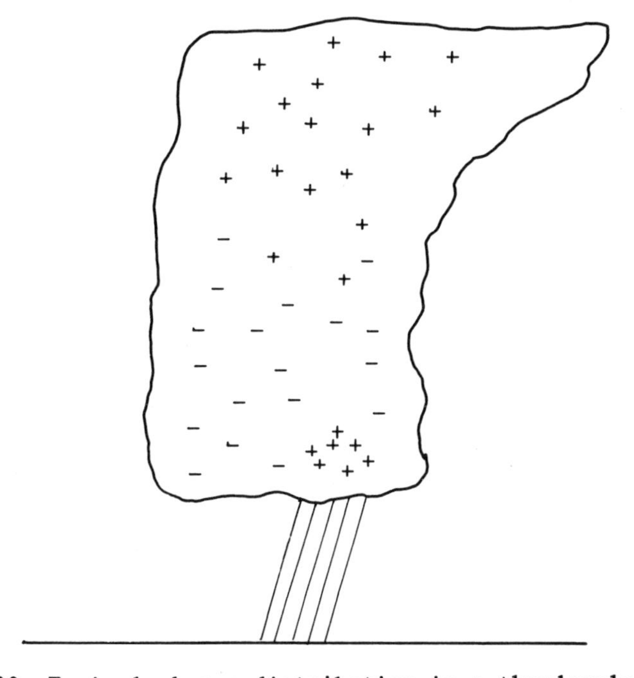

**Fig. 30 Typical charge distribution in a thundercloud.**

A typical diagram of a thunder cloud is shown in Fig. 30.

Many theories have been developed ever since Franklin's and D'Alibard's discoveries over 200 years ago. A theory must resort to some kind of energy source and most thundercloud theories seem to fall within two such categories, namely that of gravitational energy and energy provided by convection. Gravitational mechanisms rely on the gravitational potential energy available as precipitation particles fall from the upper part of a cloud to its lower region. Convection mechanisms make use of energy supplied by the intense updraft winds in thunder clouds.

Most theories such as the Wilson mechanism, influence charging, icing and ice splintering mechanisms all fall under the gravitational process. Charge is produced on precipitation particles by one of the above processes in which negative charge is preferentially attached to larger and heavier drops which fall down and separate from the smaller drops causing a large build up of the electric field within the cloud. One serious problem with gravitational mechanisms is that according to the data presented by Israel (see sec. 4.2 and 4.3) an average thunderstorm cell which contains $8 \times 10^{8}$ kg of water must produce $1.7 \times 10^{13}$ joules of electric energy, which is not possible. Simple calculations show that the gravitational energy available in a cell which contains $8 \times 10^{8}$ kg of water has a potential gravitational energy of $0.6mgh/2 = 1.2 \times 10^{13}$ joules ($g$ is the gravitational acceleration and $h$ the average vertical separation of roughly 5 km between the charge centres). This means that there is not enough gravitational energy available to account for the electric energy present in the cell. If the above reasoning holds true then all charging mechanisms which rely on gravitational potential energy can be ruled out. However, more careful evaluation of existing data and more measurements are needed before such a decision can be made.

Convection mechanisms on the other hand can easily provide

the energy and charge needed through the continuous ventilation of near stationary drops. There are essentially only two theories developed so far which are based on convection, the theory by Grenet (1947) and the electrochemical mechanism (Wåhlin 1973). Grenet and later Vonnegut (1955) suggested that natural positive space charge is brought in from below the cloud and carried to the top by convective updraft currents. The positive charge at the top of the cloud will attract negative ions from the outside air volume which are captured by down draft currents and thus assumed to be brought to the bottom of the cloud. As the process continues, the increased electric field will produce more positive ions under the cloud by corona point discharges at the earth's surface which in turn are carried through to the top by updrafts. One objection (Chalmers, 1957) is that it is difficult to understand how the updraft and downdraft currents can distinguish between negative and positive ionsinsuch a way that only negative ions can be transported by down drafts and positive ions by updrafts.

In the electrochemical charging mechanism convection currents simply bring natural ions of both signs in from outside and ventilate them through the drop population of the cloud. However, negative ions which are extremely reactive attach themselves to drop surfaces at the bottom of the cloud while the inert positive ions sieve through and are carried by the updrafts to the upper regions. The electrochemical charging mechanism, which is described in Chapter 3, is very effective and despite criticism by Vonnegut, Moore, Griffith and Willet (see section 2.7.3) is the only process which can easily be modelled in the laboratory. The diagram in Fig. 31 shows such a model constructed from two clumps of steel wool. The two sections of steel wool, one representing the bottom and the other the top of a thunder cloud, are electrically insulated from each other. A fan blows air through the cloud model to simulate the updraft winds inside a thunder cloud. A radioactive alpha particle source (Po 210, 500 micro Cu) near the fan feeds positive and negative ions at equal number into the cloud model. Negative ions are immediately captured by the electrochemical process in the bottom section while the

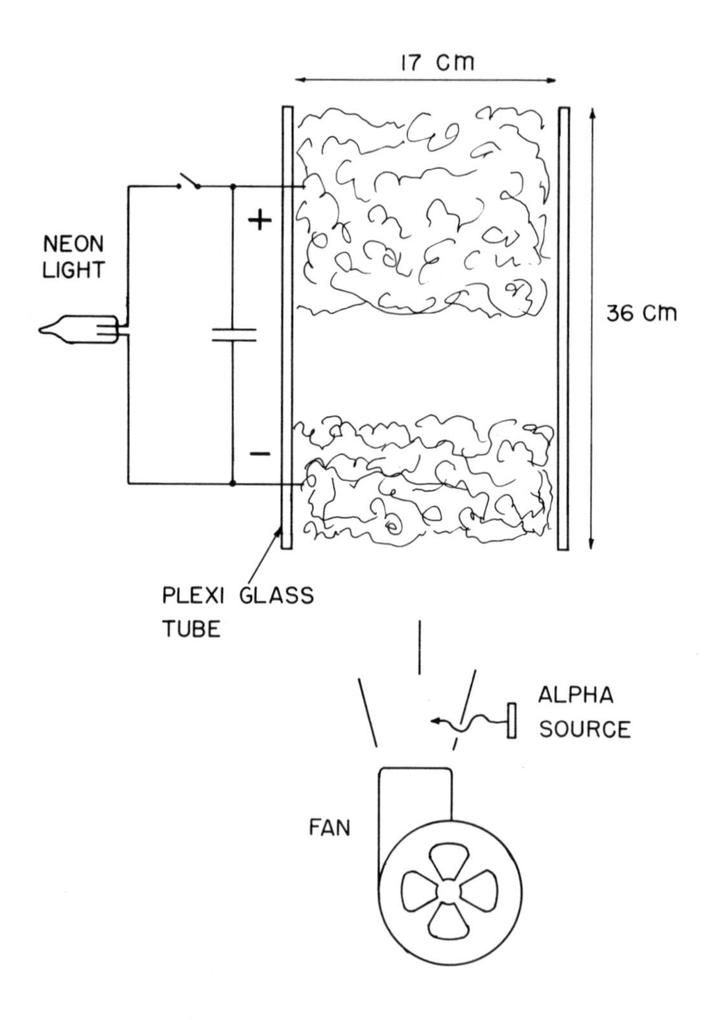


**Fig. 31 Laboratory model of thundercloud.**

inert positive ions pass through and reach the upper steel wool section where they stop together with the air flow. In a model the size of a large football over one hundred volts of charge separation is readily achieved. A small neon light, connected across the cloud model, demonstrates the lightning discharge in a realistic manner. The polonium 210 is readily available from several mail order suppliers and does not require a licence.

The electrochemical charge separation in clouds has been theoretically scrutinized by Pathak, Rai and Varshneya (1980) who determined that enough charge can be separated from atmospheric ions within 10 minutes to trigger the first lightning discharge. Once lightning has occurred numerous ion pairs are created by the discharge itself which will increase the charge rate drastically to allow for ensuing lightning discharges to occur within seconds of each other. In the dissipating stage when down drafts begin to appear in an organized way, the charging process can reverse and form a small positive charge centre near bottom of the cloud. This explains the small pocket of positive charge near the cloud base which is often observed by investigators (see Fig. 30).

### 4.7.1 BALL LIGHTNING

Ball lightning is a controversial subject and such a rare phenomenon that many investigators in the field still doubt its existence. Many sightings have been reported dating back hundreds of years, but no convincing photographs or scientific records have ever been produced. Nevertheless, ball lightning has intrigued a great number of reputable scientists and has resulted in many exotic theories. Certain fundamental characteristics of ball lightning have been agreed on, which can be described as follows: The geometrical shape of ball lightning is that of spheroid with a diameter of 10 - 20 cm although diameters as small as 2 cm and as large as 150 cm have been reported. The colour of ball lightning varies from white, bluish-white, greenish-white, red or orange-red. The light or glow seems to be steady throughout its duration which might be a few seconds to several minutes. Ball lightning might decay slowly and faint out or disappear in a sudden explosion often causing severe damage. Ball lightning has been reported to pass through window panes and walls, a phenomenon which is difficult to conceive. In most cases it moves silently above ground or along fence wires and power lines. Occasionally it is accompanied by small sparks and corona discharges which indicate the presence of electrostatic charges. Ball lightning appears only during severe thunderstorms that produce intense lightning discharges. It is believed that ball lightning might originate from the ionized lightning channel of super bolts.

Small electric plasma balls have often been observed in submarines when battery banks have been accidentally shorted out. The high short-circuit currents cause intense arcs to appear across contacts of reverse current relays which are fitted with magnetic blow-out coils. Occasionally a fire ball is blown out which will float off into the engine room and remain visible for a short time while decaying by the normal electron-ion recombination process.

As previously mentioned, there are many theories proposed in order to explain ball lightning (Barry, 1980). One such theory is the ring current plasmoid model which is derived from observations and experiments performed with submarine batteries (see Silberg, 1965). The ring current model does not mention what mechanism creates a ball lightning but deals more with the decay time of the plasma itself. Other more exotic theories are the standing wave model which is based on microwave resonance and the crystal model which treats the problem from a quantum mechanical point of view.

### 4.7.2 THE QUANTUM MODEL

The quantum model is interesting in that it tries to explain how a completely ionized plasma is maintained at very low temperatures, less than 340°C. Neugebauer (1937) used quantum mechanical arguments, derived from well-known free-electron gas models, to show that once a plasma ball has been created it will require very little energy to sustain itself. For example, from the black body radiation of such a plasma it can be determined that a completely ionized ball of 10 cm in diameter would require less than 24 watts and a 20 cm ball about 90 watts to remain completely ionized. The quantum model does not explain where the power comes from that generates the heat or how the plasma ball was created in the first place.

### 4.7.3 THE STANDING WAVE MODEL

Cerrillo (1943) and Kapista (1955) advanced a model of ball lightning in which the ionized plasma is produced and maintained by radio frequency oscillations. It is believed that a standing wave is created inside a plasma where the size of the plasma is equal to one quarter wavelength of its cavity frequency. If the radio frequency is suddenly cut off, the ball will produce a weak shock wave when collapsing. If the RF energy is gradually diminishing then the ball will slowly faint away without any drastic after effects. How to explain the mechanism that is responsible for producing such a high intensity RF field remains a problem.

### 4.7.4 THE RING CURRENT MODEL

Fire balls formed in engine rooms of submarines have led to the speculation that ring currents might be responsible for the production of plasmoids with reasonably long life times. The long life time is based on the idea that a plasmoid must have a very low ohmic resistance $R$ and a certain amount of induction $L$, due to the magnetic field of the ring current. The time constant $t = L/R$ is therefore a measure of the plasmoids life time. Fig. 32 shows a diagram of the ring current model with its surrounding magnetic field. However, the power required to maintain the plasma must come from the stored magnetic field itself. The problem is that most of the magnetic field is of no use because it is situated outside the plasma ring and very little energy is trapped inside the

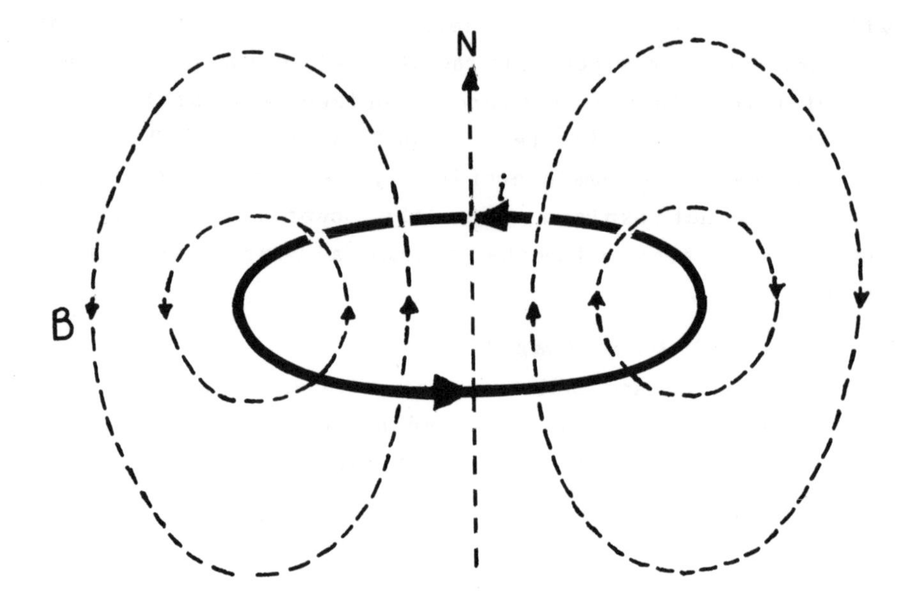


Fig. 32 The ring current model of ball lightning.

plasmoid itself. Also, it is not clear how such a ring current might be produced and how it can obtain a spherical shape.

### 4.7.5 THE PINCH EFFECT

The pinch effect is a well known phenomenon. It explains how lightning bolts sometimes break up into small balls called

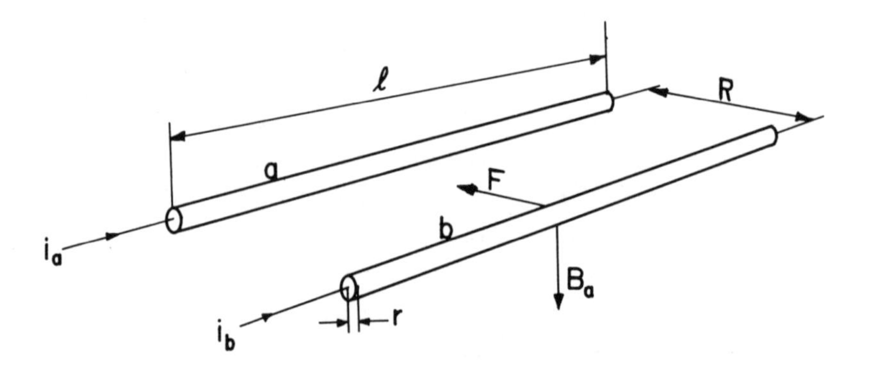


Fig. 33 Two parallel wires which carry currents in the same direction attract each other.

bead lightning or pearl lightning, an event which has been observed by the author in both Africa and Sweden. The strong magnetic field produced by the lightning bolt current can force the flow of charge to bend inward towards the centre of the lightning channel to such an extent that it will choke and pinch itself off. The pinch effect can be better understood if one examines Ampere's law, see Fig. 33, which states that the attractive force between two parallel conductors, which carry currents $i'$ and $i''$ respectively, is equal to

$$F = l \cdot \frac{i' \, i'' \, \mu_0}{2\pi R} = l \, i' B'' , \tag{13}$$

where $l$ is the length of the current elements which are separated by a distance $R$. The magnetic field generated by one current at the site of the other is $B$ and $\mu_0$ is the

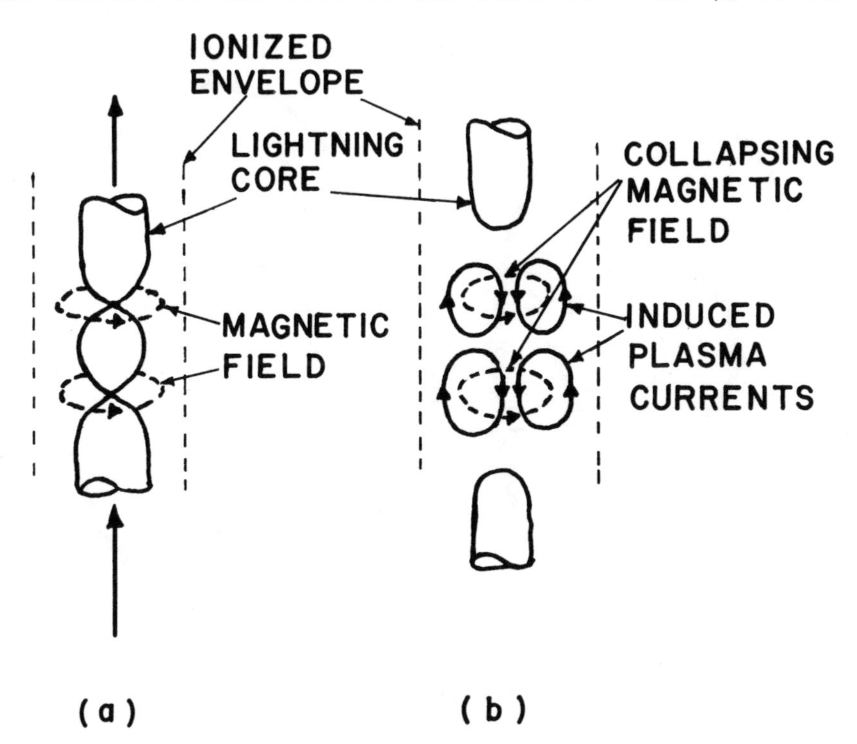


Fig. 34 The pinch effect causes the lightning channel to break up into small segments (bead or pearl lightning).

permeability constant. In a lightning channel, where numerous charges or current elements move in the same direction, a mutual attractive force is developed which will cause the lightning channel to contract, see Fig. 34 a. The force of contraction can reach a magnitude where the lightning channel gets pinched off and becomes separated into small spherical sections (beads or pearls) see Fig 34 b. Once the flow of charge is pinched off, the electric current will cease which causes the magnetic field to collapse. The collapse of the magnetic field will induce eddy currents in the ionized plasma, isolated by the electromagnetic pinch effect, which will maintain the electric discharge and prevent the magnetic field from rapid decay. The plasma ball formed can be pictured as a magnetic ring field trapped inside a toroid electric plasma current, see Fig. 35. One of the conditions

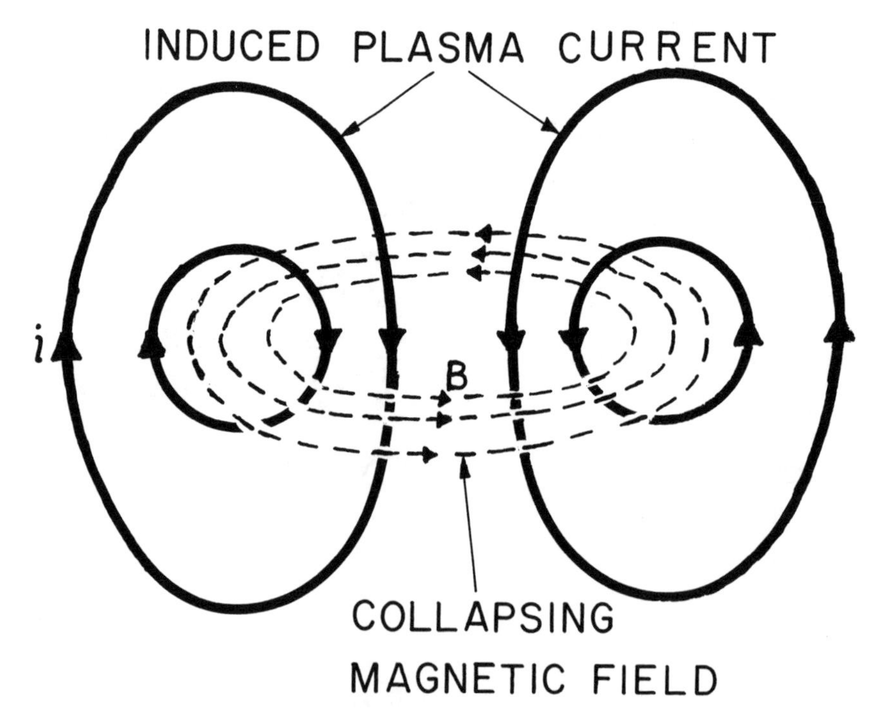


**Fig. 35 Magnetic ring field model of ball lightning.**

required is that a plasma must be completely ionized and superconducting in order to offer minimal resistance or ohmic loss. Any presence of neutral particles will cause recombination and rapid decay. A collision with a conductive object such as an iron stove, for example could drastically neutralize the plasma ball in an explosive manner. The theoretical model described implies that a ball 10 cm in diameter, completely ionized, would contain an ionization energy of 7 kilo joules and eight times that amount for a 20 cm diameter plasma ball. The field of induction set up by the plasma current is

$$B = \mu_0 I \ , \tag{14}$$

where $I$ is equal to the lightning current at the moment of pinch-off. The energy density of the trapped magnetic field is

$$U_B = \tfrac{1}{2} \frac{B^2}{\mu_0} \ . \tag{15}$$

The energy of the magnetic field in a 20 cm diameter ball, produced by a super bolt delivering 400 kA peak current, is equal to about 400 joules. Using Neugebauer's argument that 20 watts is sufficient to maintain a 20 cm plasma ball (see 4.7.2) means that 400 joules will make the plasma ball last for about 20 seconds.

# CHAPTER 5
# Fairweather Phenomena

## 5.1 EXPERIMENTS AND OBSERVATIONS

After Lemonnier and Beccaria discovered the fairweather electric field by probing the atmosphere with poles and stretched wires, the first considerable advance was made by de Saussure (1779) who introduced several new methods of measurements. De Saussure constructed an electrometer consisting of two elder pith balls, suspended from silver wires inside a glass jar with metallic shielding. When the balls were charged to the same polarity they repelled one another and the amount of deflection was a measure of the electric intensity from a source brought in contact with the silver wires. By raising and lowering an antenna which was connected to the electrometer, de Saussure could show that the electric potential in the atmosphere increased with altitude. Another clever experiment by de Saussure was to throw a lead ball, which was connected to the electrometer by a thin wire, straight up in the air, breaking the wire and electric connection. The ball, after reaching a certain potential, would take its charge with it and leave an opposite charge behind on the electrometer. De Saussure realized that the electrostatic charge induced on his antennas was caused by a positive static charge in the atmosphere above. He is also believed to have been the first to discover the systematic annual change in the fair weather field, see section 1.5.

One interesting experiment involving the use of kites was performed by Mahlon Loomis, a Washington D.C. dentist, during the period 1862 - 1873. Loomis used two kites at the end of stranded bronze wires about 600 feet in length. The wires were both grounded through sensitive galvanometers but the ground connections could be broken by means of telegraph keys. Loomis was able to transmit morse code messages from one antenna to the other over a distance of up to 16 miles without the use of any external electric source. In the early 1870's Loomis was supposed to have achieved voice communication using a Berliner's microphone. He applied for government funds for a classified military project, but since the Civil War was over, Congress would not appropriate the money. What is important is that Loomis must have been first to achieve wireless communication but obviously not by radio but by disturbing the electric fairweather field. A member of the family, Thomas Appleby, has published a book on Loomis' experiments and copies of patents issued (Appleby 1967).

In the beginning of this century it was discovered that ions cause the atmosphere to be conductive and new theories were needed to explain how charges on the earth-atmosphere system are maintained. It has been suggested that the earth's rotation in its own magnetic field could produce a magneto effect which would drive the electric fairweather circuit, or that evaporation of water from the earth's surface might charge the earth negatively and the water vapour positively by some unknown process. In fact, Volta (1800) advanced an electrification theory based on vaporization and condensation of water in order to explain charge production in thunderclouds. Today, although the question still remains open, there are primarily two theories dealing with the problem of the origin of the electric fairweather field. One is the closed circuit hypothesis which asserts that all thunderstorms around the world recharge the earth-ionosphere system. The other theory is the electrochemical process which is an electrostatic mechanism based on the preferential capture of negative ions by the earth's surface. The amount of charge produced by all thunderstorms around the earth at any given time is certainly adequate for replenishing the fairweather charge and the variations of the fairweather field as a function of GMT seems, to a certain extent to follow the variation in the world-wide convection pattern as shown in Fig. 15. The electrochemical charging mechanism, on the other hand, suggests that if a sphere the size of our earth is ventilated by ionized air then, by extending the curve in Figure 23, it will charge to approximately one million coulombs, a value that conforms with the measured charge on earth. The curve, of course, shows the charge on water surfaces, but it should be remembered that two-thirds of the earth is covered by water. Also, the earth's surface is constantly being ventilated by winds carrying natural ions which are supplied at a steady rate by cosmic and radioactive decay. The excess positive ions left behind near the earth's surface are carried aloft by convection and by eddy diffusion, thereby mixing through the atmosphere to heights of several kilometres. Lifting the positive ions from near ground to higher altitudes stretches the electric field lines and increases the potential. This is analogous to the contact potential experiments performed by Lord Kelvin (see section 3.2) in which the potential is increased by separating the two metal plates in his condenser apparatus. The electrochemical charging mechanism is strongly dependent on convection and is also expected to follow the world-wide convection pattern as shown in Fig. 15.

About 90% of the electric fairweather field is confined to altitudes of less than two kilometres, or that part of the atmosphere referred to by meteorologists as "the mixing region". Integrating electric field strengths up to a variety of heights will give the values of electric potentials at varying altitudes. At an altitude of two kilometres, for example, this potential is normally about +200 kilovolts with respect to earth. At the University of West Virginia a team of physicists have managed to build an electric motor driven by the electric fairweather field utilizing an antenna extended high into the atmosphere (Jefimenko, 1971).

Since the electric fairweather field is enhanced by convection and mixing in the fine weather atmosphere, measurements of electric field strength at varying altitudes and at diverse locations may give important information about the atmospheric mixing processes in terms of local weather forecasting. Large scale forecasts might also be possible because the electric fairweather field exhibits a slight variation every twenty-four hours that is synchronous for all parts of the world. The diurnal fluctuation is linked to the fact that continental land masses of various sizes are exposed by the daily rotation of the earth to the heat of the sun and this results in convection and enhancement of the electric field (see Fig. 15).

Many other phenomena in the atmosphere can be explained by the electrochemical effect. For example, airplanes in flight are subject to ventilation by atmospheric ions and thus charge accordingly. A Boeing 747 should reach a potential of over -100 kV by rough estimation based on Equation (7). A large body such as a zeppelin could reach a potential of several hundred thousand volts in the naturally ionized atmosphere, making a voyage in a hydrogen airship a highly dangerous venture, as proved by the mysterious disintegration of the Hindenburg in 1937.

Charging also occurs inside combustion chambers of engines where ions are formed at high temperatures. By means of the electrochemical process, ions of one sign (normally negative) are preferentially captured by engine walls, permitting those of opposite charge to escape in the exhaust stream. Exhaust charging is a common phenomenon in our modern world. Not only do aircraft charge by this process but motor vehicles on rubber tyres become electrostatically charged even at stand still while idling. Exhaust gas from engines and smoke stacks produce excessive positive space charge over heavily populated areas. Aircraft flying through such areas during landing approaches become positively charged in contrast to what is normally experienced during regular fairweather flights. The explanation is that if an aluminium fuselage is subject to ventilation of an abnormally high positive to negative ion ratio, where $N^+/N^- = 1.3$ or more then, according to Fig. 11, the aircraft must charge positively.

Exhaust charging also plays an important role in geophysical phenomena such as volcanic eruptions which often produce lightning and during earthquakes where a glow or earthquake light is emitted along fissures and cracks.

It is difficult to say whether atmospheric electricity has any practical applications at the present time. The structure of the fairweather field and how it is linked to convection and turbulence in the atmosphere might be of meteorological interest. The fairweather field as an energy source is too feeble for practical purposes and will only yield about 1μwatt per $m^2$ at the earth's surface, or approximately one billion times less than solar power.

Also worthy of consideration is the fact that for the past 50 years scientists have been looking for physiological effects caused by ions and electric fields in the earth's atmosphere. Many experiments have already been performed that seem to show such a relationship. However, one still lacks concrete physical evidence and much more work is needed before one can determine the significance of atmospheric electricity with respect to one's state of health. There is also a growing interest in planetary atmospheric electricity as a consequence of data sent back from interplanetary space probes.

### 5.2.1 EXHAUST CHARGING

Lord Kelvin was aware of the "wonderful agency in flames and fumes" causing deviations in his electrometer readings when passing through his condenser plates. Earlier (c. 1800) Volta discovered that charges on conductors changed rapidly when exposed to candle flames or lit fuses. Modern science explains that ions produced from combustion increase the conductivity of the surrounding air causing potential gradients to change more rapidly. An increase in ion population also enhances the electrochemical charging effect. In combustion chambers of engines, where numerous ions are formed, the electrochemical charging effect often becomes a nuisance and in some cases a hazard. For example, fuel might ignite while attempting to refuel a charged vehicle or aircraft. During combustion negative ions are usually captured by the engine walls while the positive ions are carried away by the exhaust stream. The result is an accumulation of negative charge on the vehicle. Fig. 36 shows an experiment performed by the author with a diesel car blocked up on insulators. A charging current of 2 micro amperes was registered at fast idle of the 2.25 litre engine. Similar experiments were carried

**Fig. 36 Exhaust charging of a motor car.**

out on small aircraft and negative charging currents of up to 10 micro amperes were measured. Exhaust charging can easily produce static potentials of several thousand volts. Severe electric shocks can be obtained from large aircraft not fitted with static eliminators or by touching metal hooks lowered from hovering helicopters.

### 5.2.2 VOLCANIC ERUPTIONS

Volcanic plumes charge to high potentials and, just like thunderstorms, are accompanied by a display of lightning discharges. Not much is known about the charging mechanism involved and very few measurements are available that can shed light on the electric structure of such plumes. Space charge measurements at ground level near volcanic eruptions reveal negative charge on precipitating ash particles which, according to the mirror-image effect (sec. 2.5), means that there is positive charge on the plume above. Opposite signs have been reported and the conclusion is that ash particles from different volcanoes charge differently. It is difficult if not impossible to determine the dominating sign of charge in plumes from ground measurements because sign reversals, due to lightning discharges and mirror-image effects, are always present.

It has been suggested that volcanic plumes charge by frictional effects between small and large ash particles and that separation of charge is enhanced when larger particles fall out of the plume. Other charging effects to be considered are convection mechanisms and exhaust charging. Convection and condensation of water vapour as in regular thunderstorms can also explain the build-up of charge in the plume. Exhaust charging might play a dominating role since numerous ions must be present in the hot interior of the volcano. Negative ions can be expected to be adsorbed before reaching the earth's surface while the more inert positive ions are carried up with the plume, see Fig. 37.

Electric measurements during volcanic eruptions are obviously very difficult to perform if not dangerous. Also, volcanic eruptions are not as common as ordinary thunderstorms and data is, therefore, very sparse.

Fig. 37 Electrostaic charging of volcanic plume.

### 5.2.3 EARTHQUAKE LIGHT

Earthquake light is a rare phenomenon and its existence is still being questioned by some scientists. It was not before 1965, during the Matsushiro earthquakes in Japan, that actual photos were obtained of earthquake light (Yasui, 1968 and 1971). Earthquake light has been observed as a faint red and white glow for hundreds of miles around just before the earth begins to shake and crack along a fault zone. There has been much speculation on what might cause earthquake light. One fact is certain, that the light must come from atoms that have been ionized or excited by some mechanism triggered by the earthquake. The light is emitted when a detached or excited electron falls back again into its stable orbit around the atomic nucleus. Ionizing collisions in air, between atoms and particles or between atoms and photons, by far outnumber exciting collisions. Ionization of air molecules, to the extent where the sky will glow for miles around, could be caused by corona discharges near the earth's surface or by ionized exhaust gas escaping from the hot interior below the earth's crust. It has been suggested that frictional heating of a shear zone will occur during an earthquake and that the frictional heat will lead to vaporization of water in and near the shear zone (Lockner *et al*, 1979). The result would be a drastic increase in the electric resistivity of the rock throughout the shear zone coupled with some sort of charge separation due to the evaporation of water. After enough charge has been collected in the shear zone, corona discharges along its top edge would stream into the atmosphere like St. Elmo's fire. One problem, however, is that corona discharges draw a great deal of current (about 100$\mu$A per cm$^2$) and it is difficult to understand how such a current can be supplied over a highly insulating rock surface. Also, corona is quite a noisy phenomenon unlike silent earthquake light.

Other theories take into account the piezoelectric effect in which strong electric fields are presumed to form in quartz rich crustal rocks subject to high mechanical pressures (Mitzutani *et al* 1976). Again high electric fields along the fault zone would create corona and St. Elmo's fire. One interesting idea is that the sparks and crackle from piezoelectric discharges in the interior would create electromagnetic radiation, similar to that of old spark transmitters used before the invention of vacuum tubes and transistors, to the extent that the air above the earth's surface would ionize and glow. The power of the electromagnetic radiation must be high enough to ionize air and make it glow. One would expect that such a powerful electromagnetic radiation, that will cause electric breakdown of air, would cook everything else in sight and heat the earth's surface to extreme temperatures.

Perhaps the explanation of earthquake light has nothing to do with electrostatic charging but simply the escape of gases such as the radioactive element radon. According to the *Handbook of Chemistry and Physics* there is about 1 gramme of Radium per square mile of soil to a depth of 6 inches which releases radon into the air. During an earthquake, radon is released from considerable depths and brought to the earth's surface in large amounts. Each radon atom is capable of producing one million ion pairs. The handbook states that when radon is frozen below its melting point it will exhibit a brilliant phosphorescence which becomes yellow as the temperature is lowered and orange-red at the temperature of liquid air. The light is from ionizing collisions between alpha particles and atoms. It can be expected that when large quantities of radon escape it might be accompanied by other foreign gases trapped far beneath the earth's surface. The gases, unnoticed by humans, could easily be detected by animals and the unfamiliar scent might have a disturbing effect on their behaviour. It is believed that the Chinese are able to predict earthquakes and it is well known that they are monitoring animal behaviour for this purpose.

## 5.3 BIOLOGICAL EFFECTS

The exposure of human and animals to ions and strong electric fields has only very recently become an important issue in many countries. The Soviet Union, for example, has very rigorous rules and regulations concerning the exposure of workers to electric fields in power stations. In Sweden studies have been carried out on the effects of ions and strong electric fields on humans under different working environment (Lövstrand *et al* 1978 and Backman, 1979). Many other countries are and have been involved in similar research. The exposure to electric fields is in itself not considered important, but the production and movement of ions in electric fields are of chief concern. Ions carry charge to the body; they are also inhaled and trapped in the respiratory system. The question is; do ions affect our physiological or psychological well being? So far, numerous tests have been performed in different environments where ion concentrations and electric fields have been changed under controlled conditions. ECG and EEG recordings, blood and urine tests, as well as blood pressure and neurasticthenics (headache, nausea, insomnia, tiredness) have been monitored by many investigators but no conclusive results have ever been reported. There are many who believe that negative ions affect them in a pleasant way and that an excess of positive ions has an adverse effect. The author knows of physicists who installed negative ion generators in their cars and which are turned on on long trips in the event their wives and children become rowdy. One atmospheric electrician confided that occasionally he turns on a positive ion generator, which puts his wife to sleep, so that he can watch a TV programme of his own choice!

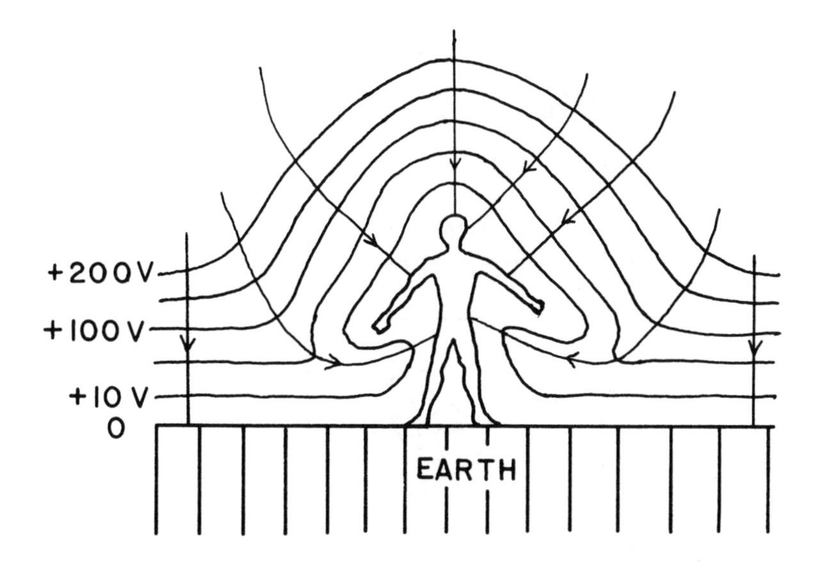


**Fig. 38 A person exposed to the electric fairweather field.**

It has been established that ion currents produced on the human body as a consequence of the electrostatic fields, are too insignificant to cause any noticeable effects (Backman, 1979). A person exposed to the normal fairweather field, see Fig. 38, will perhaps collect a positive ion current of $10^{-13}$ A on his head. Internal brain currents are millions of times larger and it is hard to believe that the external ion current will have any effect on the function of the brain. The effect of inhaled positive and negative ions on the

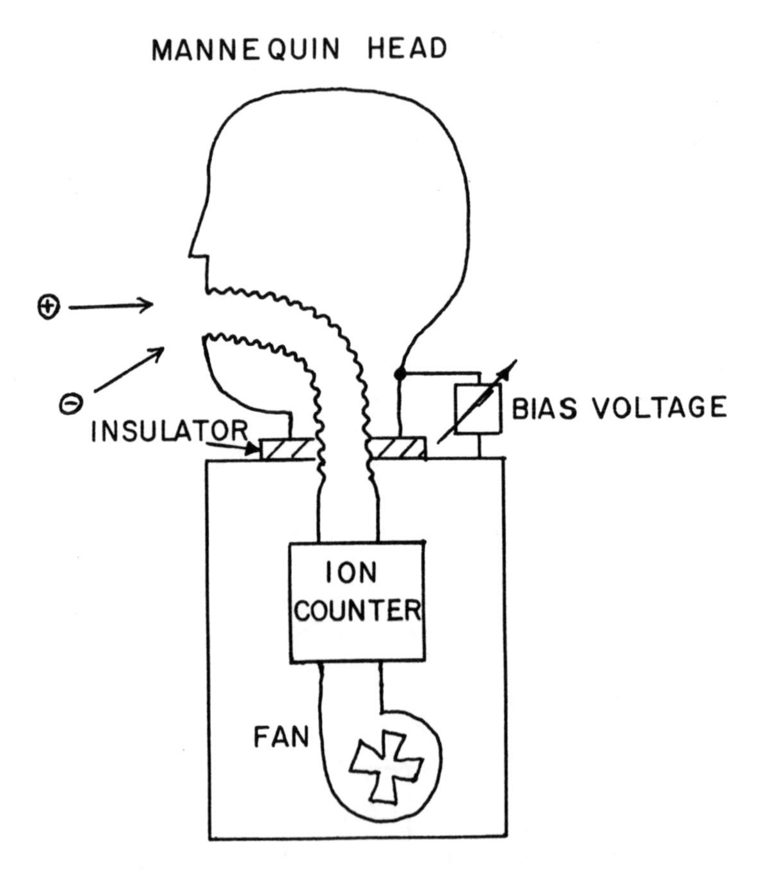


Fig. 39 Apparatus measuring ion intake through the respiratory system.

respiratory system is unknown. It is possible that negative

ions might have a stimulating effect on the cilia in the alveoli. The amount of ions inhaled and the sign, whether negative or positive, is strongly dependent on the electrostatic charge carried by the person in question. For example, a person walking across a carpet in an office might charge negatively to several hundred volts. The strong electric field will repel negative ions and allow only positive ions to enter the respiratory system. In fact Bach (1963) has demonstrated this effect using an insulated mannequin head biased at different electric potentials, see Fig. 39. The result is shown in Fig. 40 where N' represents the increase in either positive or negative ions as compared to the normal ion ratio at zero bias.

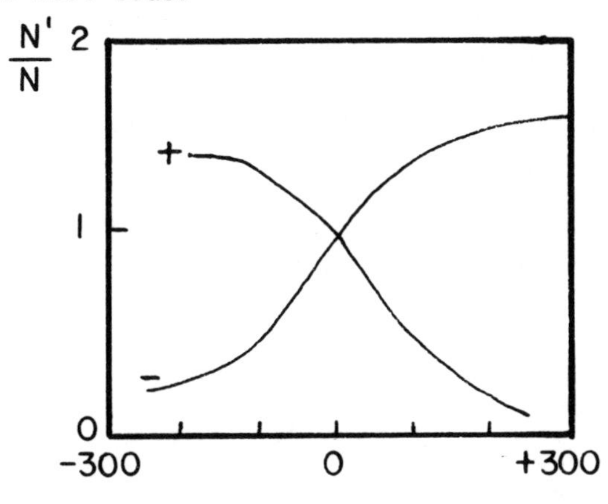


Fig. 40 Ratio of charged ions inhaled at different bias voltages.

The belief that ions might have a bioelectric effect on humans and animals probably started in the Alps of Europe. It has always been noticed that during the föhn when dry winds sweep down the mountainside some people become depressed or develop head-aches and that dogs and chicken tend to hide away. It has also been observed that the positive space charge or positive ion concentration in the atmosphere during the föhn is very high which is probably one reason why positive ions are blamed for causing discomfort amongst mountain villagers. Another observation is that warm mountain winds, like the föhn, are very dry which can cause considerable discomfort, as explained by Tromp (1969).

Warm mountain winds occur when cold air masses pass over mountain ranges, and when the cold heavy air reaches the other side it falls down, compresses, and heats up again. When cold air warms up it is capable of holding more water vapour and its relative humidity decreases. The warm dry air will take up water from the surrounding environment and dry everything in sight. The Indians in the Rocky Mountain region call the dry mountain wind "the chinook", which means snow-eater, because of the rapid disappearance or vaporization of the snow in its path.

The high positive ion concentration can be explained by the fact that winds ventilating the other side of the mountain lose negative ions which are depleted by the electrochemical effect while the excess inert positive ions will follow the air stream over the top, see the diagram in Fig. 41. As the

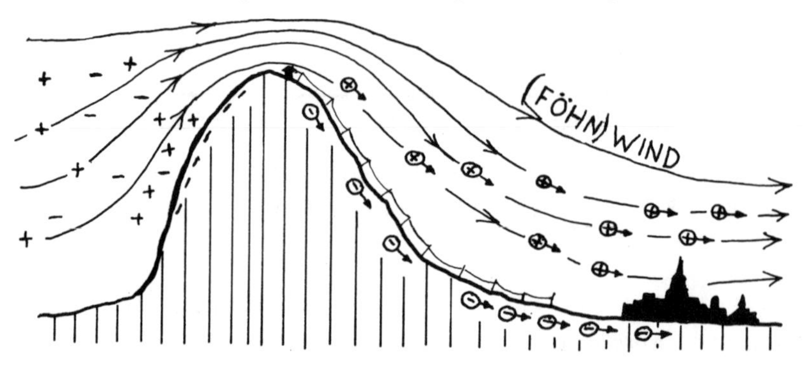


**Fig. 41 Dry mountain winds (föhn or chinook) and positive ions.**

positive space charge moves down the mountain side it is followed by an induced negative image charge along the earth's surface. The induced negative charge is particularly noticeable on telegraph lines going up and down the mountain side. A protruding telegraph line will carry induced negative charge down the mountain which is registered as an electric current in the line flowing up the mountain.

## 5.4 THE ELECTRODE EFFECT

The electrostatic fairweather field near the earth's surface exhibits some very peculiar variations ascribed the so called electrode effect. For example, the electric fairweather field can be slightly higher just a few metres above the earth's surface than at the earth's surface itself. This is due to a sudden increase of negative ions in a thin layer above the earth's surface. From Fig. 12, section 2.2, it was explained that the fairweather field can be thought of as electric field lines which connect the excess positive ion population in the atmosphere with trapped negative charges on the earth's surface. If, however, a cloud of negative ions was suddenly

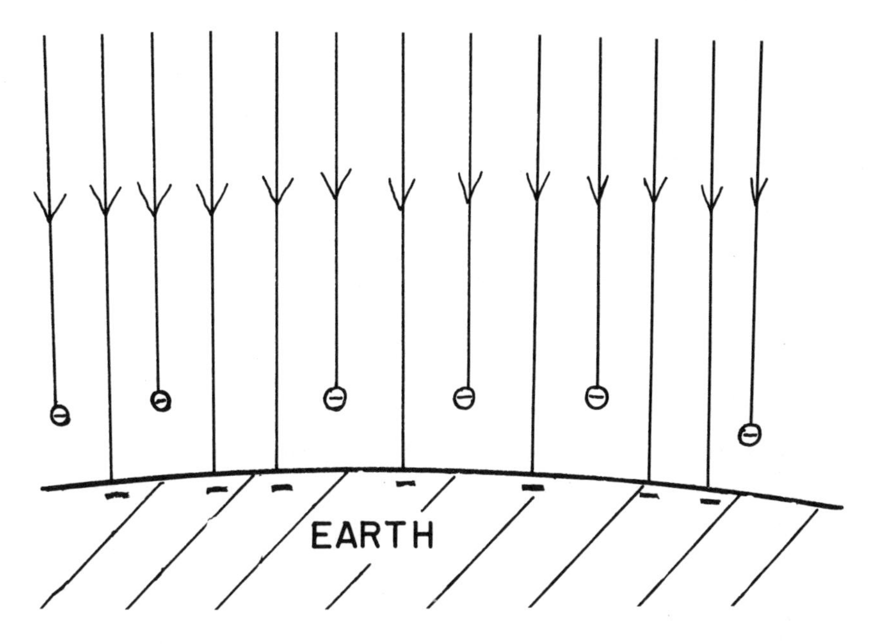


Fig. 42 The electrode effect.

formed near the earth's surface it would mean that some of the field lines in Fig. 12 do not reach ground but will terminate on those negative ions. The number of field lines per unit surface area would, therefore, be less at the surface and higher just above the thin negative ion cloud, see Fig. 42. A sudden increase in negative ion concentration near the earth's surface could be caused by the passage of a cloud above with a strong positive charge centre. The passage of a positive charge above would temporarily force positive ions to the earth's surface leaving an excess of negative ions behind in a thin layer above ground. The reverse effect is also possible when a negative charge centre passes overhead in which case there is a considerably stronger than normal electric field very near the earth's surface. It is also believed that an electrode effect exists even in the absence of clouds. The idea is that, since positive ions congregate towards the earth's surface because of the normal fairweather field and since there is no supply of negative ions leaving the surface, an imbalance must occur with a higher than normal positive ion concentration near the earth's surface. Numerous measurements, however, do not clearly show such an effect.

## 5.5 INTERPLANETARY STATIC ELECTRICITY

That static electricity is present on other planets has been established by instrumentation carried on board unmanned space probes. For example, lightning discharges have been observed on Jupiter by the Voyager spacecraft (Smith *et al*, 1979, Cook *et al*, 1979 and Gurnett *et al*, 1979). There is also a strong belief that lightning might occur on Venus.

The rate of current carried by cosmic ray particles to the earth is about 0.2 amperes. This means that the earth will charge 280 volts per second positively unless there is a flow of electrons to the earth at the same rate. The electrons might come from the sun, in which case the sun must charge positively. The existence of strong electrostatic fields in interplanetary space is an untouched field of research. There are problems with orbiting spacecraft building up electrostatic charges which can have adverse effects on measuring instruments and data collected. Extreme charging might occur on space vehicles when returning to earth through the highly ionized layers of the ionosphere. The ventilation of ionospheric ions in combination with the electrochemical charging effect might cause the plasma discharge observed during re-entry. This makes radio communication impossible.

# CHAPTER 6

# Instrumentation and Measurements

## 6.1 MEASUREMENTS

Atmospheric electricity is a field that is very easy to get into because it does not require a large capital investment for measuring equipment. It is a difficult field, however, when it comes to the understanding and interpretation of the various measurements. For example, despite vast data collected over two centuries, there is still no agreement on how thunderclouds charge or how the earth-atmosphere fairweather field is maintained.

Today's advanced technology offers eloquent and accurate instruments with fast response time. It is now possible to construct simple electrometers and electromagnetic radiation detectors using chip circuitry. Video recording equipment is becoming a household item that also has great potential for lightning research, making it possible to view flights of lightning bolts in micro-second intervals.

## 6.2 ELECTRIC FIELD MEASUREMENTS

Most common are measurements of the electric field at or near the earth's surface. The potential gradient at the earth's surface varies in many ways but usually averages about +100 volts per metre during fine weather in the absence of clouds. Except for the normal diurnal variations of the fairweather field (see sec. 2.2), strong rapid changes are often registered which might be due to electrically charged dust or heavy pollution. Marked changes in the fairweather field have been observed before the onset of fog or during temperature inversions in the lower atmosphere. Many of the above field effects and their causes are still not fully understood. The electric field below thunderclouds is usually very strong and easy to detect. Rapid field changes occur during lightning discharges and field reversals are common. Field measurements below thunderclouds reveal such features as charging rates, electric polarity and discharge rates, to mention a few.

There are several ways one can measure the electric field in the atmosphere. A simple approach is to connect an electrometer to an antenna of some sort such as shown in Fig. 43. The antenna can be a conducting disc placed at a

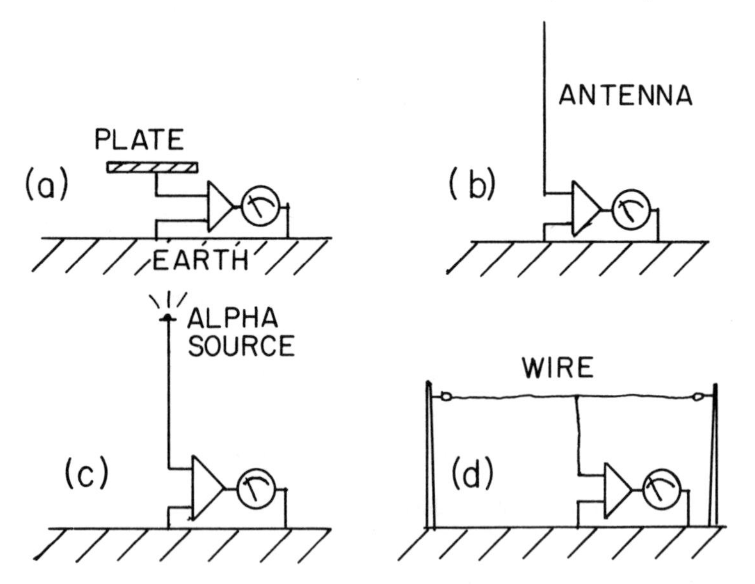


Fig. 43 Different antenna configurations for electric field measurements in the atmosphere.
(a) Disc antenna.
(b) Whip antenna.
(c) Whip antenna with radioactive probe.
(d) Long wire antenna

predetermined height above ground (Fig. 43a). The disc will

charge and reach a potential which is equal to or very near the potential of the atmosphere at that height. The diagram in Fig. 43b shows an electrometer connected to a short whip antenna which gives voltage readings that are difficult to calibrate since the antenna whip will protrude through many levels of electric potential. An ion-producing radioactive alpha source at the tip of the whip antenna, see Fig. 43c, will increase the electric conductivity in the air near the tip and ensure a better accuracy of potential measurement as a function of height. A long wire, suspended above ground at predetermined levels, will give accurate readings of atmospheric potentials as a function of height, see Fig. 43d.

## 6.3 THE ELECTROMETER

There are many excellent electrometers commercially available, but since the electronic market of today can offer a variety of sophisticated integrated circuits, it makes simple

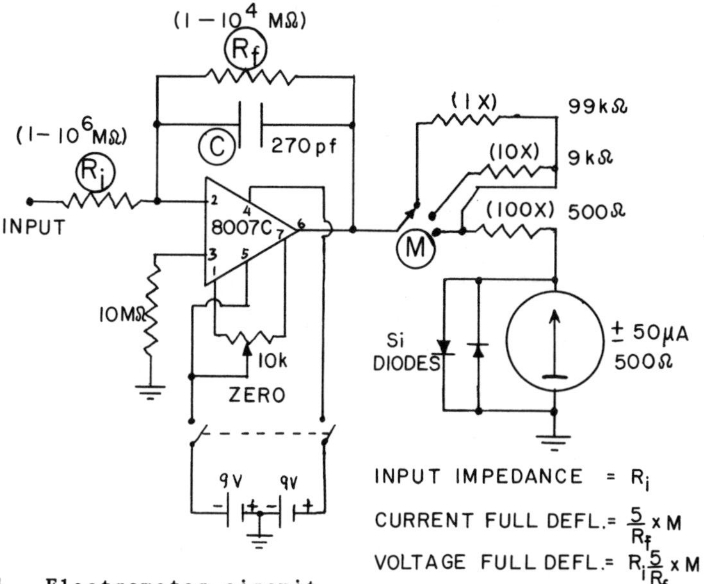


Fig. 44 Electrometer circuit.

home made devices nearly as effective. For example, the circuit in Fig. 44 has been used by the author in many experiments including the Gerdien cylinder measurements described in section 2.1. It uses two 9 volt batteries and is portable, which makes it convenient for experiments in the field. The electrometer is mounted in an electrically shielded metal box.

Measuring electric fields and field changes during thunderstorms will produce many interesting results. Lightning discharges cause rapid field changes and polarity reversals which can be registered on a chart recorder. The recovery time of the electric regeneration process in clouds is also a very interesting feature that can be studied with the help of a recorder. During thunderstorm measurements it is recommended that the electrometer in Fig. 44 be fitted with an $R$ $C$ network which increases the time constant of the circuit to several seconds. Because of the strong fields produced by thunderstorms, it is also advisable to decrease the sensitivity of the circuit, by changing $R_i$ $R_f$. A typical

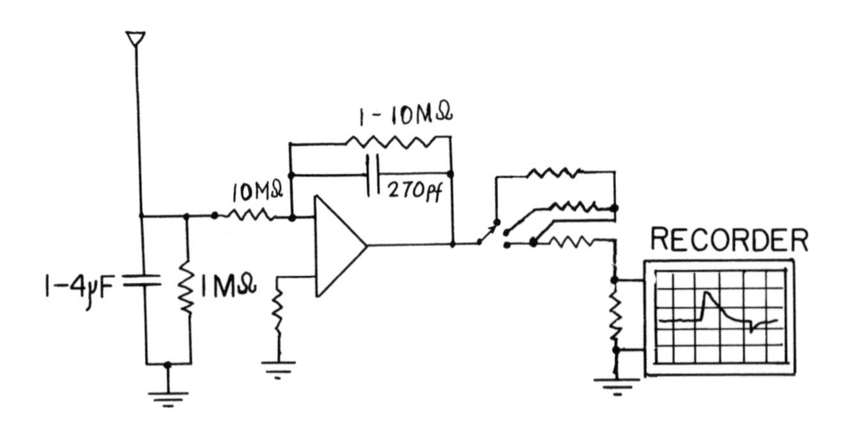


Fig. 45 Circuit for electric field measurements of thunderclouds.

circuit modification for thunderstorm measurements is shown in Fig. 45.

**6.4 THE FIELD MILL**

The electric field mill is a device based on electrostatic induction. It consists of one or two electrodes which either rotate in an electrostatic field or become periodically exposed to a field by rotating vanes. Fig. 46 illustrates a

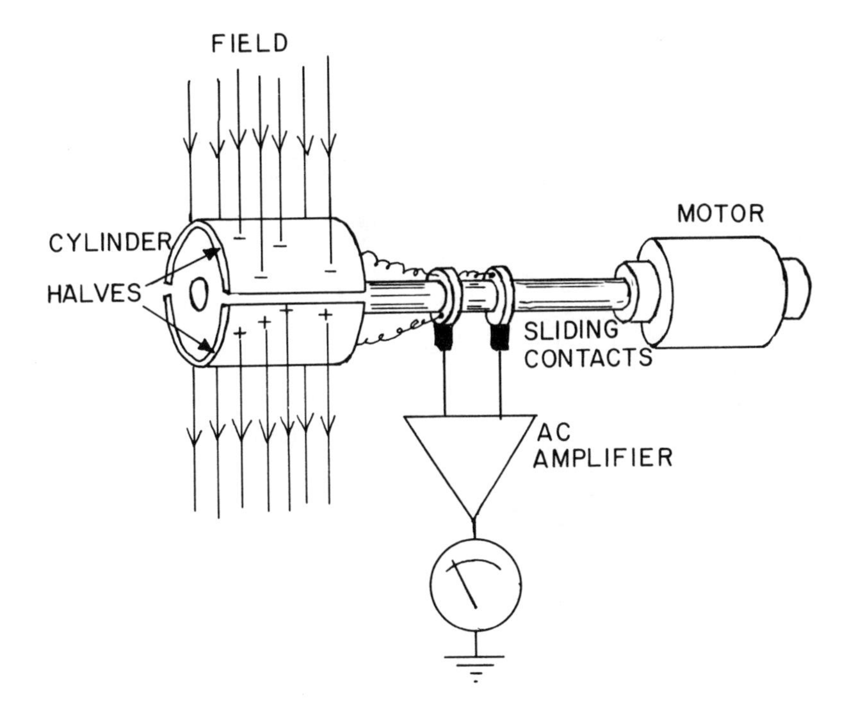


**Fig. 46 Cylindrical electric field mill.**

cylindrical field mill which consists of two cylinder halves that are electrically insulated from each other. An electric motor rotates the two halves in the electric field to be measured so that they become alternately exposed to both the positive and negative direction of the field. The result is that an alternating (ac) signal is generated across the two halves which can be easily amplified. The rotating shutter field mill, on the other hand, comprises a stationary electrode, which becomes periodically exposed to the external

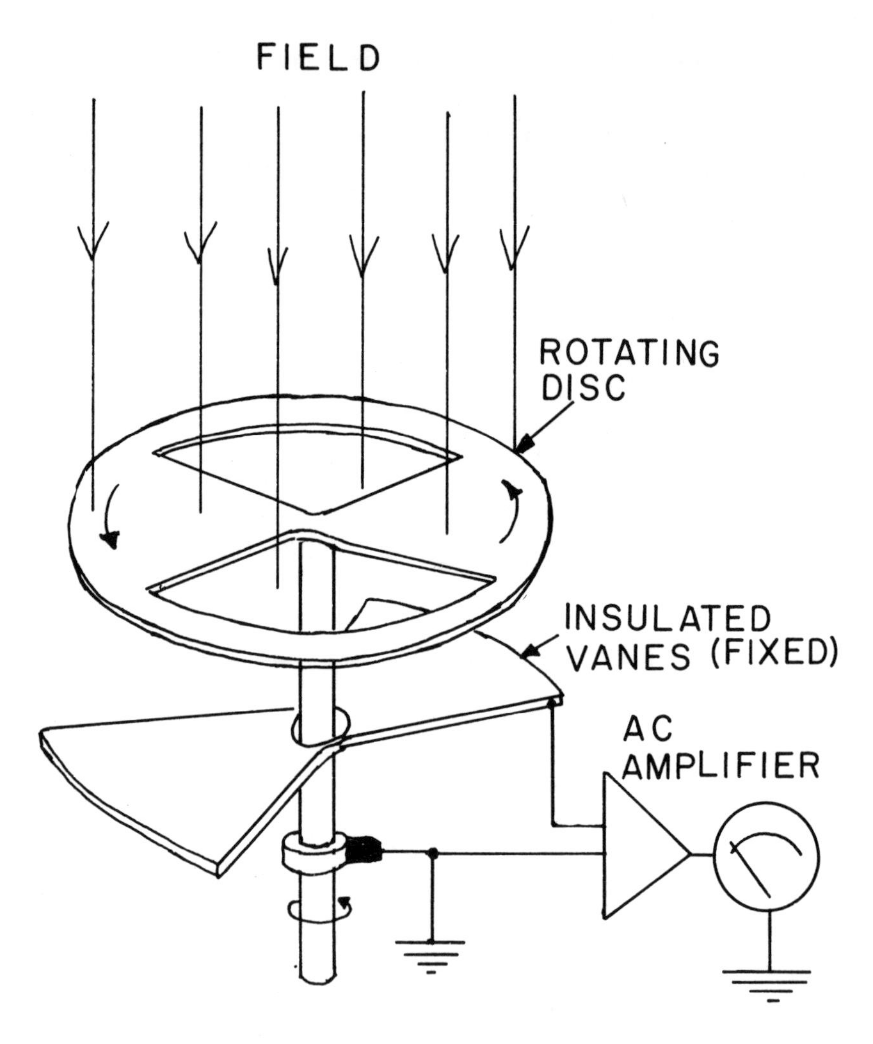


Fig. 47 Rotating shutter electric field mill.

electric field through a rotating grounded disc (see Fig. 47). A variation on this type of field mill is a stationary grounded cover plate with a rotating disc electrode. Although not commonly used, the cylindrical field mill has the advantage that when mounted in a fixed position it can also indicate the direction of the field. This is accomplished by measuring the phase shift of the ac signal relative to the orientation of the two cylinder halves. The electric field mill is widely used in both atmospheric research and in industry where electric fields need to be measured. The rotating shutter type field mill has to be pointed towards the source of the field in order to obtain a maximum reading. This feature is useful when searching for static charges that might cause problems in the laboratory or industrial plant. The rotating shutter field mill is commercially available in several countries.

### 6.5 ELECTROMAGNETIC DETECTION

The crackling sound of atmospherics generated by lightning is familiar to everyone who listens to radio. "Sferics" interference is usually stronger in the low frequency band (10 - 500 kHz) were the electromagnetic energy spectrum of lightning is most powerful. There is also an interest in the higher frequency bands of lightning radiation, around 10 - 30 MHz, relating to corona discharges and fast field changes. An attractive feature of atmospherics is that it enables long distance detection of lightning storms. Radio receivers with directional antennae are used to determine the position and movement of thunderstorms. Several stations can accurately pin-point storms by triangulation. Loop or frame antennae, such as seen on ships, are used and a typical electric block diagram is shown in Fig. 48. Experience has shown that the shape of electromagnetic pulses received from lightning changes with distance, which makes it possible to estimate the distance from the antenna to the discharge. Some aircraft are fitted with "stormscopes" utilizing the above principles to determine the direction and distance of nearby intracloud discharges in order to avoid dangerous confrontation with lightning. Radar echoes of lightning bolts can be seen as reflections superimposed on echoes of heavy precipitation in clouds. It is believed that the ionized channel, produced by lightning, is large enough to cause reflection of radio waves. Radar observations of thunderstorms bring out many interesting

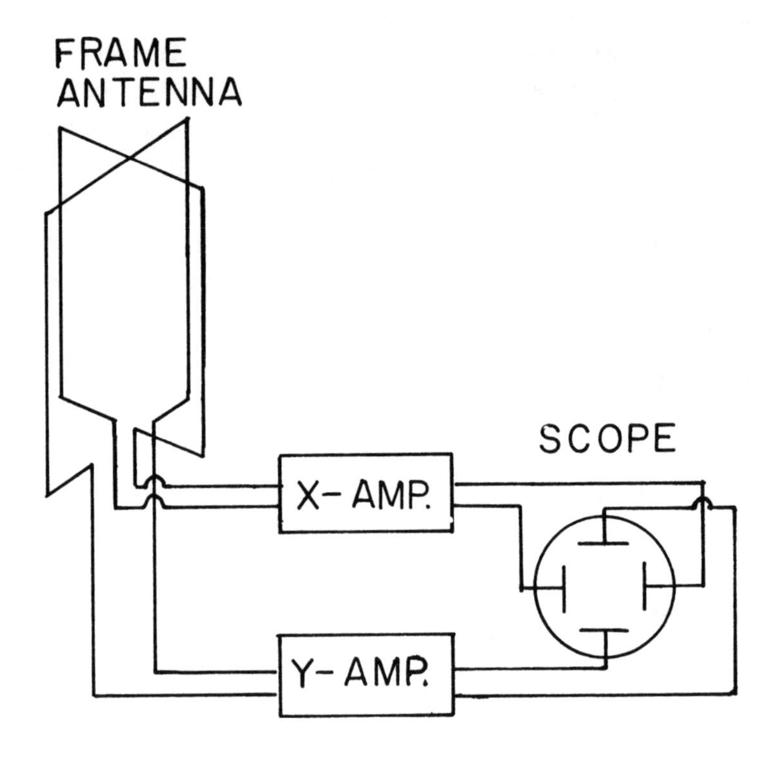


Fig. 48 Loop antenna for lightning detection.

features which might aid in the understanding of the physical processes at work in clouds. For example, radar reflections from precipitation appear several minutes before any visible electrical activity. The radar reflections rise at first together with the cloud top but start to descend as soon as lightning occurs. At the same time lightning commences, precipitation becomes visible and starts to fall out of the cloud. One question often asked is whether lightning induces gushes of precipitation; or *vice versa*.

## 6.6 THE FUTURE

There are still many problems outstanding in the field of atmospheric electrostatics. New and better measuring devices will help solve some of the problems still at large, but equally important is the development of new ideas and techniques.

Electrostatic fields and ions in the atmosphere and their effect on our well being is still an open question. There are no conclusive results yet available and more research is needed. It is known, however, that both positive and negative ions reduce the life of bacteria (Serratia marsescens) residing in aerosols and that high negative ion concentrations seem to offer relief to persons suffering from bronchial disorders.

Static electricity in the atmosphere might not offer much practical value but it has intrigued many of the best theoreticians and experimentalists throughout time. There is no shortage of thunderstorms and lightning but the problem is how to safely penetrate a cloud and study its micro structure. One alternative is to use large cloud chambers capable of producing convective clouds. Some of the problems might be solved with the aid of orbiting space laboratories or unmanned satellites. The world wide thunderstorm activity and lightning distribution can be monitored from space and correlated with electric field measurements at earth. This could help settle the very important question whether or not the fairweather field is generated by thunderstorms. Another approach to the above problem is the use of computer modelling techniques. Electric field plots of thunderclouds could determine the amount of current, if any, delivered to the ionosphere. The number of field lines that extend from the top of a cloud to the ionosphere, compared to the number that reaches ground, indicates the fraction of current delivered by the cloud to the fairweather circuit. The previous title in this series *Computer Modelling in Electrostatics* by D. McAllister, J.R. Smith and N.J. Diserens deals in detail with such modelling techniques.

# References


Abbé Nollet, 1749, Guerin, Paris.

Appleby, T. 1967, *Mahlon Loomis Inventor of Radio*, Loomis Publications, Washington, D.C.

Bach, C. 1963, *Det luft-elektriske klima. Rosenkilde og Bagger*, Köpenhamn.

Backman, C.M. 1979, Report from Institute of High Voltage, University of Uppsala, UURIE 79:116.

Barry, J.D. 1980, *Ball Lightning and Bead Lightning*, Plennum Press, New York.

Bruce, C.E.R. and R.H. Golde, 1942, J. Instn. Electr. Engrs.**88**, 487.

Beccaria, G.B. 1775, Turin.

Cerrillo, M. 1943, *Comision Impulsora y Coordinatora de la Investigacion Cientific Mex. Ann.* **1**, 151.

Chalmers, J.A. and R. Gunn, 1954, J. Met. **11**.

Chalmers, J.A. 1956, J. Atmosph. Terr. Phys. **9**, 451.

Chalmers, J.A. 1967, *Atmospheric Electricity*, 2nd ed. Pergamon Press, New York.

Cook, A.F. *et al*, 1979, Nature, **280**.

D'Alibard, T.F. 1752, Letter to Acad. de Sci. Paris, le 13 mai.

Elster, J. and H. Geitel, 1885, Ann. Phys, Chem. **25**, 121.

Elster, J. and H. Geitel, 1899, Phys. Z. **1**, 245.

Gerdien, H. 1905, Phys. Z. **6**, 647.

Gray, S. 1735, Phil. Trans., **39**. 400, 2.

Grenet, G. 1947, Ann. Geophys. **3**, 306.

Griffiths and Vonnegut, 1975, Private communication.

Gunn, R. 1957, Electrification of Aerosols by Ionic Diffusion **25**, 542.

Gurnett, D.A. *et al*, 1979, Geophys. Res. Letters, **6**, 511.

Heaviside, O. 1902, Encycl. Brit. 10th edn. **33**, 213.

Hodges, D.B. 1954, Proc. Phys. Soc. Lond. B, **67**, 582.

Hoffman, K. 1923, Beitr. Phys. Frei. Atmos. **11**, 1.

Imyanitov, I.M. and E.V. Chubarina, 1967, *Electricity of the Free Atmosphere.* Available from U.S. Dept. of Commerce, Clearing House for Federal Sci. and Techn. Inf. Springfield, VA 22151.

Israel, H. 1954, Mitt. dt. Wetterd. No. 7.

Israel, H. 1970, *Atmospheric Electricity*, TT 67-51394/1, U.S. Dept. of Commerce, Springfield, VA 22151, Vol.1.

Israel H. 1973, *Atmospheric Electricity*, TT 67-51394/2 U.S. Dept. of Commerce, Springfield, VA 22151, Vol.2.

Jefimenko, O. 1971, Amer. J. Phys. **39**, 776.

Kapista, P.L. 1955, Dokl. Akad. Nauk. SSSR, **101**, 345.

Kasemir, H. 1950, Arch. Met. Wien, A, **3**, 84.

Kasemir, H. 1950b, *Das Gewitter* ed. H. Israel, Akad. Verlag, Leipzig.

Kasemir, H. 1951, J. Atmos. Terr. Phys. **2**, 32.

Kasemir, H. 1965, *Problems in Atmosph. and Space Electr.* Ed. S.C. Coroniti, Elsevier Publ. Co. Amsterdam, 215.

Kelvin, Lord, 1860, Roy. Instn. Lecture. 208-226.

Kelvin, Lord, 1898, Phil. Mag. S.5 Vol. **46**. No. 278, 82.

Lemonnier, L.G. 1752, Memoires de l'Academie des Sciences, **2**, 233.

Levin, Z. and W.D. Scott, 1975, J. Geophys. Res. **80**, 3918.

Levin, Z. 1983, *Proceedings in Atmospheric Electricity*, ed. Ruhnke and Latham, A. DEEPAK publishing, Hampton, Virginia.

Lockner, D.A. 1979, A Mechanism for Generating Earthquake Lights, USGS, Menlo Park, CA. 94025

Loeb, L. 1958, *Static Electrification*, Springer-Verlag, Berlin.

Lodge, O. 1885, Phil. Mag. October, 383.
Lövstrand, K.G. 1978, Report from Institute of High Voltage Research, University of Uppsala, UURIE:114-79.

Lundquist, A.G. 1969, *The Lightning Stone and its History*, (in Swedish) Available from The Institute of High Voltage Research, Uppsala, Sweden.

Lutz, C.W. 1939, Gerl. Beitr. ys. Geophys. **54**, 337.

Maclean and Goto, 1890, Phil. Mag. Aug.

Mitsutani, H. 1976, Geophys. Res. Lett. **3**, 365.

Moore and Vonnegut, 1977, *Lightning*. ed. R.H. Golde, Academic Press, London, 88

Mühleisen, R. 1953, Z. Geophys. **29**, 142.

Müller-Hillebrand, D. 1963, *Torbern Bergman as a Lightning Scientist*. Available from The Institute of High Voltage Research, Uppsala, Sweden.

Neugebauer, H.N. 1937, Z. Physik, **106**, 474.

Papoular, R. 1965, *Electric Phenomena in Gases*, Iliffe Book Ltd., London, 95 and 96.

Pathak, P.P., J. Rai and N.C. Varshneya, 1980, Ann. Geophys. t.36, fasc.4, 613.

Phillips, B.B. and R. Gunn, 1954, J. Met. **11**

Reynolds, S.E. 1954, Compendium of thunderstorm electricity. U.S. Signal Corps Research Report, Socorro, N.M.

Rosenberg, A.G. 1708, *Versuch einer Erklärung von den Ursachen der Electricität*, Korn, Breslau.

Sartor, D. 1954, J. Met. **11**, 91.

de Saussure, H.B. 1779, *Voyages dans les Alpes*. Geneva.

Schonland, B.F. 1938, IV. Proc. Roy. Soc. A, **164**, 132.

Schonland, B.F. 1950, *The Flight of Thunderbolts*, Claredon Press, Oxford.

Schonland, B.F. 153, Proc. Roy. Soc. A, **220**, 25.

Silberg, 1965, *Problems of Atmospheric and Space Electricity*, ed. S.C. Coroniti, Elsevier Publ. Co. Amsterdam, 430.

Simpson, G.C. 1905, Phil. Trans. A, **205**, 61.

Simpson, G.C. 1910, Phil. Mag. **19**,, 715.

Simpson, G.C. and F.J. Scrase, 1937, Proc.R.Soc.London **161** 309.

Smith, B.A. *et al*, 1979, Science, **204**, 951.

Takahashi, T. 1973, J. Atmos. Sci. **30**, 249.

Tromp, S.W. 1969, *Der Einfluss von Wetter und Klima auf den Menschen Umschau in Wissenschaft und Technik*, Frankfurt/M, **24**, 753.

Turman, B.N. 1978, J. Geophys. Res. **83**, C10, 5019.

Turman, B.N. and B.C. Edgar, 1982, J. Geophys. Res. **87**, C2, 1191.

Wåhlin. L. 1973, Found. of Phys. **3**, 459.

Wåhlin, L. and H. Kasemir, 1985, J. of Electrostatics, **16**, 379.

Wait, G.R. 1950, Arch. Met. Wien, A, **3**, 70.

Wall, W. 1708, Phil. Trans., **26**, 69, 2.

Whipple, F.J.W. 1932, Terr. Magn. Atmos. Elect. **37**, 355.

Willett, J. 1980, Private communication.

Wilson, C.T.R. 1906, Proc. Camb. Phil. Soc. **13**, 363.

Wilson, C.T.R. 1916, Proc. Roy. Soc. A, **92**, 555.

Wilson, C.T.R. 1920, Phil. Trans. A, **221**, 73.

Wilson, C.T.R. 1925, Phil. Trans. A, **221**, 73.

Wilson. C.T.R. 1929, J. Franklin Inst. **208**, 1.

Winkler, J.H. 1746, *Die Stärke der elektrischen Kraft.* Leipzig. 1746.

Workman, E.J. and S.E. Reynolds, 1950, Phys. Rev. **78**, 254.

Workman, E.J. and S.E. Reynolds, 1953, *Thunderstorm Electricity* 139.

Wormel, T.W. 1953, Quart. J. R. Met. Soc. **79**, 3.

Yasui, Y. 1968, Proc. Kakioka Magnetic Observ. **13**, 23.

Yasui, Y. 1971, Proc. Kakioka Magnetic Observ. **14**, 67.

# Index